IMPORTANT

P9-CJH-947

MCGRAW-HILL
OURCES

HERE IS YOUR REGISTRATION CODE TO ACCESS
YOUR PREMIUM McGRAW-HILL ONLINE RESOURCES.

For key premium online resources you need THIS CODE to gain access. Once the code is entered, you will be able to use the Web resources for the length of your course.

If your course is using **WebCT** or **Blackboard**, you'll be able to use this code to access the McGraw-Hill content within your instructor's online course.

Access is provided if you have purchased a new book. If the registration code is missing from this book, the registration screen on our Website, and within your WebCT or Blackboard course, will tell you how to obtain your new code.

Registering for McGraw-Hill Online Resources

TO gain access to your MCGraw-Hill web
resources simply follow the steps below:

1. USE YOUR WEB BROWSER TO GO TO: **http://www.digitalzoology.com**

2. CLICK ON **FIRST TIME USER**.

3. ENTER THE REGISTRATION CODE* PRINTED ON THE TEAR-OFF BOOKMARK ON THE RIGHT.

4. AFTER YOU HAVE ENTERED YOUR REGISTRATION CODE, CLICK **REGISTER**.

5. FOLLOW THE INSTRUCTIONS TO SET-UP YOUR PERSONAL UserID AND PASSWORD.

6. WRITE YOUR UserID AND PASSWORD DOWN FOR FUTURE REFERENCE.
   KEEP IT IN A SAFE PLACE.

TO GAIN ACCESS to the McGraw-Hill content in your instructor's **WebCT** or **Blackboard** course simply log in to the course with the UserID and Password provided by your instructor. Enter the registration code exactly as it appears in the box to the right when prompted by the system. You will only need to use the code the first time you click on McGraw-Hill content.

Thank you, and welcome
to your MCGraw-Hill
online Resources!

* YOUR REGISTRATION CODE CAN BE USED ONLY ONCE TO ESTABLISH ACCESS. IT IS NOT TRANSFERABLE.
   0-07-248952-9 HOUSEMAN, DIGITAL ZOOLOGY VERSION 2.0

REGISTRATION CODE

anxiolytic-47628173

Mc
Graw
Hill **Higher
Education**

Digital Zoology
Version 2.0
CD-ROM
and
Student Workbook

Jon G. Houseman
University of Ottawa

Visit www.digitalzoology.com for interactive crossword puzzles
and answers to text questions and self-quizzes from this workbook.

Boston   Burr Ridge, IL   Dubuque, IA   Madison, WI   New York   San Francisco   St. Louis
Bangkok   Bogotá   Caracas   Kuala Lumpur   Lisbon   London   Madrid   Mexico City
Milan   Montreal   New Delhi   Santiago   Seoul   Singapore   Sydney   Taipei   Toronto

Windows 95, 98, NT, ME, 2000

Program Requirements

In order for Digital Zoology 2.0 to function properly, you must have the following programs installed on your machine:

>*Flash Player version 5 or greater
>*Internet Explorer 5 or greater

Installing the Required Programs

Run the installer by double-clicking on the DigitalZoology.exe file, which is located at the top level of the CD. The installer will check your system for the above programs and prompt you to install them if they are not already installed. After running the installer, a shortcut to the *Digital Zoology* 2.0 program will be added to your desktop.

>Please close all other programs before running the installer. If you should encounter any problems with the installation, restart your computer and run the installer again.

Running Digital Zoology 2.0

To run the program, just double-click on the desktop shortcut 'Digital Zoology V2.0'. You can also open the *Digital Zoology* 2.0 program by double-clicking on the 'Digital Zoology V2.html' shortcut located at the top level of the CD.

Macintosh Power PC, OS 8.5 or Later

Software Requirements

In order for Digital Zoology 2.0 to function properly, you must have the following programs installed on your machine:

>*Flash Player version 5 or greater
>*Internet Explorer 5 or greater

If you are unsure whether you have these programs installed, try running Digital Zoology 2.0. If you encounter problems or errors with the program, it is a good indication that you need to install some or all of the above programs.

Installing the Required Programs

Included on this CD are the installers for the required programs mentioned above. They can be found in the folder named "installers." There are two folders in the "installers" folder, one for each of the different programs you need to install.

>The Internet Explorer 5 installer can be found in the installers/ie5 folder and the Shockwave Flash installer can be found in the installers/shockwave_flash folder.

Running Digital Zoology 2.0

After you have installed all of the required applications, you can start the program by double-clicking on the file named 'Digital Zoology V2.html,' which is located on the CD.

McGraw-Hill Higher Education

*A Division of The **McGraw-Hill** Companies*

Student Workbook for use with
DIGITAL ZOOLOGY, VERSION 2.0 CD-ROM
JON G. HOUSEMAN

Publisher: Margaret J. Kemp
Senior developmental editor: Donna Nemmers
Marketing manager: Heather K. Wagner
Media producer: Renee Russian
Cover designer: David W. Hash
Cover photos: © Diane R. Nelson
Workbook photos: © Jon G. Houseman
Production supervisor: Sherry L. Kane
Supplement producer: Audrey Reiter
Copyeditor: Anne Scroggin

Published by McGraw-Hill Higher Education, an imprint of The McGraw-Hill Companies, Inc., 1221 Avenue of the Americas, New York, NY 10020. Copyright © 2003, The McGraw-Hill Companies, Inc. All rights reserved.

 2 3 4 5 6 7 8 9 0 QPD/QPD 0 3 2
ISBN 0-07-248952-9 (workbook), 0-07-256481-4 (packaged with CD)

www.digitalzoology.com

TABLE OF CONTENTS

AUTHOR'S PREFACE

About *Digital Zoology*

*D*igital Zoology has existed in a variety of different forms over the years. It started as a series of digital photos that I used in class to help bring the specimens that students saw in lab to the lecture hall. It then grew with the addition of more specimens and interactive cladograms. Soon after, it found a home on the computer networks at the university. When students wanted to take it home, it ultimately became the CD-ROM that evolved into the *Digital Zoology* Free Preview Version, published in December, 2000. By January 2001 over 15,000 students across North America were using the program. Version 1.0 followed a year later with the addition of more vertebrate specimens and an expanded interactive glossary. It's now a year later, and like any good arthropod, *Digital Zoology* has molted once again. We've added more invertebrate specimens, doubled the size of the interactive glossary, and provided you with what we call "Read Abouts," which take a detailed look at the over 100 animal groups included in *Digital Zoology*. But the biggest change is in the program. It is now an Internet, web browser-based program and you can run it from the CD-ROM included in this workbook or from the *Digital Zoology* website: www.digitalzoology.com. If you use the website you'll find additional features that aren't available on your CD-ROM: the ability to take notes and place personalized bookmarks in the program for handy reference to sections important for your studies. This workbook provides a quick look at different animals traditionally studied in zoology labs and serves as a useful study tool for review of lab dissection exercises.

What's new in Digital Zoology *Version 2.0*

- ○ Coverage of seventeen new phyla: Placozoa, Ctenophora, Cycliophora, Sipuncula, Onycophora, Tardigrada, Bryozoa, Rotifera (including Acanthocephala), Gnathostomulida, Chaetognatha, Gastrotricha, Nematomorpha, Priapulida, Kinorhyncha, Loricifera, Phoronida, and Brachiopoda. The Annelida has been updated to include the Echiura and Pogonophora. Protozoans have been reorganized based on the infrakingdoms of Actinopoda, Alveolata, Discicristata, and Sarcomastigota.
- ○ Ten new lab modules for Version 2.0: Broyzoans, Brachiopods, Leech, Sea Cucumber, Gastrotrichs, Rotifers (including Acanthocephala), Hemichordata, Protochordates, Lamprey, and a special module on insect mouthparts. Similar in style to the other lab modules in *Digital Zoology*, each new module includes a quiz. New lab modules for Hemichordata and for the chordate subphyla Urochordata and Cephalochordata replace the previous chordate origins units, providing a complete taxonomic organization to *Digital Zoology* Version 2.0.
- ○ New taxon "Read Abouts" in Version 2.0 replace the taxon boxes in Version 1.0 and include information on over 100 taxa, the number of living and extinct species, habitats, and how they function.
- ○ Updated taxonomy of the animal phyla allows for easy comparison of the differences between traditional morphological phylogenies and those created with molecular data.
- ○ The number of video clips in *Digital Zoology* Version 2.0 is almost twice that of Version 1, with forty new video sequences for clam, jellyfish, sea star, sea cucumber, sea urchin, perch, rotifer, gastrotrich, frog, shark, and more protozoans.
- ○ More than 250 glossary terms have been added, bringing the total interactive key terms in *Digital Zoology* Version 2.0 to over 750 terms. The full glossary can be consulted at any time or by clicking on highlighted key terms found throughout the program.
- ○ Image galleries are now available for many of the major phyla, providing photos that detail diversity within the phyla.

The CD-ROM and Online Components

Digital Zoology is organized around four components. The first provides illustrations, photos, and video clips of the various specimens that will help students to understand the specimens and groups of animals commonly studied in the lab. Digital Zoology contains over forty-five different species, covering a full range of organisms from microscopic amoebas to vertebrates such as the frog and fetal pig. It provides students in whole animal biology courses such as general zoology, invertebrate zoology, and vertebrate zoology with an interactive guide to the specimens and materials that they will be studying in their laboratory and lecture sessions. Each lab module contains illustrations, photographs, and annotations of the major structures of organisms and microscope slides commercially available from the suppliers used by high schools, colleges, and universities. The second component involves interactive cladograms that explore the origins and phylogeny of the various animal taxa at the phylum, subphylum, and class level. The third is synoptic "Read Abouts" that provide information on over 100 different animal groups. Special terms and text in each "Read About" are linked to an interactive glossary to help students understand the terms unique to the group. (You can browse through the full glossary by selecting "Glossary" from program tools found on the green taxon bar.) The fourth and final component of Digital Zoology is self-evaluation tools that help you assess how well you know the material. Interactive crossword puzzles test your knowledge by taking traditional one-word answer questions and organizing them into a crossword puzzle. Self-tests in each chapter of the Digital Zoology student workbook are combined with drag-and-drop quizzes in each lab module to test your knowledge of the anatomy of the different specimens you'll encounter in your zoology course.

The Student Workbook

The student workbook is designed to complement the specimens on the Digital Zoology CD-ROM and on-line. If a particular taxon isn't included in the current version of Digital Zoology, there won't be a chapter on it in the workbook. Each chapter begins with a short description of the defining differences for the group of organisms — phylum or class — covered in the chapter. These differences, many of which can be seen in the lab or in Digital Zoology, combine characteristics that define the taxon and highlight the unique ways that organisms in the group function. The workbook takes a closer look at a representative specimen for each taxon and explores the unique biology of the animal. A convenient "structures checklist" allows students to keep track of the key structures for each of the specimens they examine. Each chapter contains self-evaluation tools: crossword puzzles that test knowledge of key terms and concepts for that group, and a self-test quiz for lab exercises. For those of you who have seen previous versions of Digital Zoology, the labeling quizzes in this workbook are similar to the drag-and-drop quizzes in Digital Zoology. These quizzes are still in Digital Zoology, only there are many more of them: a quiz is now included for all the major animal groups covered.

There are a number of additional interactive elements in the student workbook. Questions in each chapter explore some of the key concepts about the group of animals discussed, and the Digital Zoology website provides answers to the questions raised in the workbook. Crossword puzzles provide a self-evaluation tool for each chapter's content, and interactive versions of the puzzles, with solutions and hints, are available on the CD-ROM or the website. An additional interactive element in each chapter of the workbook is the self-test quiz. Students can identify and label key structures in photographs of microscope slides or dissected specimens.

This workbook is not intended to replace the zoology textbook or laboratory manual. Rather, it has been designed as a supplement to both. In a large part, Digital Zoology focuses on things that students will be able to see in their lab sessions. A full-color copy of the workbook is available from the Digital Zoology website.

A Note to Students

I hope you'll find *Digital Zoology* to be a valuable tool in your studies of zoology. Here are some of the ways that my students have used the program.

Lab Preparation. *Digital Zoology* is a great way to prepare for lab sessions. Part of this preparation should always be to read through the lab manual before your lab. Unfortunately, manuals often have only a few illustrations or photos of the specimens you'll be looking at in the lab. They really don't prepare you for how things will look when you dissect your specimen. *Digital Zoology* contains many photos of the specimens at various stages in the dissection, and by working through the lab module prior to doing the lab you should be able to get a good idea of what to expect once you start your dissection.

Almost all lab courses have some sort of lab quiz where you have to look at a specimen and answer some question about it (for example: "identify the structure labeled A and what it does"). I've tried to duplicate that with the printed self-test here in the student workbook and the drag-and-drop quizzes on the *Digital Zoology* CD-ROM or on-line. Solutions to the printed self-tests are available on the *Digital Zoology* website, and the drag-and-drop quizzes in *Digital Zoology* are interactive, allowing you to find the correct answers. If you want even more practice at identifying structures, cover up the lower-right corner of the screen to hide the legend in *Digital Zoology*. Take out a piece of paper and write what you think the legend should be, using the lettered pointers on the screen. Compare your legend to the one in *Digital Zoology*.

Learning the Language. Zoology, like any discipline, has a set of terms, or lexicon, all its own. The text in *Digital Zoology* may appear to be full of words that don't have any meaning to you as you begin your course, but you'll know the terms by the end of it. In *Digital Zoology*, key words and terms are highlighted in blue. As you read the text in the taxon information boxes and within the interactive cladograms, resist the urge to click on a highlighted term right away. Instead, take a scrap of paper and write out a short definition for the term. Then, click on the term and compare your definition to the one in *Digital Zoology*. A full glossary is available in the student workbook and in the electronic version of the program in Program Tools.

One of my students' favorite activities are crossword puzzles, and so I've provided paper versions in the workbook and interactive versions in *Digital Zoology*. The clues in the crossword puzzles are similar to the type of one-word answers, or fill-in-the-blank type of questions, that often appear on exams and quizzes. Each puzzle has between thirty and forty different clues to test your knowledge on each of the different groups of animals.

Finally, throughout the workbook you'll find some short-answer questions on the animals in each chapter. You'll find answers to these questions on the *Digital Zoology* website.

A Note to Instructors

Many of the photographs and illustrations in *Digital Zoology* are available for you to use in your class. You can download PowerPoint slides from the *Digital Zoology* website for each of the major taxa covered in *Digital Zoology*. Feel free to modify, alter, or customize the slides to meet your requirements. The design and color scheme of the slides has been optimized so the images can be used as 35mm slides, transparencies, or digital projection. If you find a photograph or illustration in *Digital Zoology* that isn't included in the PowerPoint slides, all of the images in *Digital Zoology* are available on the web.

The images used in *Digital Zoology* are part of the large BIODIDAC image base (biodidac.bio.uottawa.ca) created for biology teachers to use in their courses. BIODIDAC is funded in part by Heritage Canada, a ministry of the Government of Canada. The image bank consists of over 3,000 photographs, illustrations, annotated drawings, and video clips of biological materials. All the material in the BIODIDAC image bank has been donated by biology teachers and is available royalty- and copyright-free for the creation of noncommercial teaching materials. BIODIDAC is always looking for additional materials, and if you would like to contribute you can find out how by visiting the BIODIDAC website.

I'd like to take this opportunity to share with you some observations on how *Digital Zoology* has changed my lectures and labs over the past ten years of its use. The ability to bring photos of the same materials students see in the lab to the lecture provides a clear link that binds these two components of the course that much closer together. Our analysis of student use of *Digital Zoology* has shown us that most of the students go through the lab module prior to doing the lab. Students tell me that the photos help them to anticipate what their dissection is going to look like during its various stages. We have a few computers in the lab with *Digital Zoology* running, and as students dissect they look for a visual concordance between their work and the modules. *Digital Zoology* wasn't designed to replace the dissection of specimens; instead it was my hope that students would gain a better understanding of the dissection through the use of the program. The crossword puzzles have been a tremendous success with my students. The clues in the puzzles are similar to fill-in-the-blank questions, and all the clues are available at the instructors' website for *Digital Zoology*.

Purchasing Options

Digital Zoology Version 2.0 can be purchased individually or packaged with other McGraw-Hill textbooks at a discounted price. Contact your McGraw-Hill sales representative for details.

A Selected Bibliography

The following resources were invaluable in preparing the *Digital Zoology* Version 2.0 and the student workbook.

Barnes, R. S. K. 1984. *A Synoptic Classification of Living Organisms*. Blackwell Scientific Publishers.

Brusca, R. C. and G. J. Brusca. 1990. *Invertebrates*. Sinauer Associates Inc.

Grassé, P. P. *Traité de zoologie: anatomie, systématique, biologie*. Masson. A multivolume set published between 1948 and 1977.

Hallan, J. 2000. *Synopsis of the Described Taxa of the Living World*. http://entowww.tamu.edu/research/collection/hallan/all1.txt

Hickman, C. P., L. S. Roberts, and A. Larson. 2001. *Integrated Principles of Zoology*. 11th edition. McGraw-Hill Higher Education.

Kardong, K. V. 1997. *Vertebrates: Comparative Anatomy, Function, Evolution*. 2nd edition. WCB/McGraw-Hill.

Margulis, L. 1998. *Five Kingdoms: An Illustrated Guide to the Phyla of Life on Earth*. 3rd edition. Freeman.

Moyle, P. B. and J.J. Cech. 2000. *Fishes: An Introduction to Ichthyology*. 4th edition. Prentice Hall.

Nielsen, C. 2001. *Animal Evolution: Interrelationships of the Living Phyla*. 2nd edition. Oxford University Press.

Parker, S. P. (ed.) 1982. *Synopsis and Classification of Living Organisms*. McGraw-Hill.

Pechenik, J. A. 2000. *Biology of the Invertebrates*. 4th edition. McGraw-Hill Higher Education.

Pough, F. H., R. M. Andrews, J. E. Cadle, M. L. Crump, A. H. Savitzky, and K. D. Wells. 2001. *Herpetology*. Prentice Hall.

Tudge, C. 2000. *The Variety of Life: A Survey and a Celebration of all the Creatures that have Ever Lived.* Oxford University Press.

Walker, W. F. and K. F. Liem. 1994. *Functional Anatomy of the Vertebrates: An Evolutionary Perspective.* Saunders College Publishing.

Willmer, P. 1990. *Invertebrate Relationships*. Cambridge University Press.

A Special Thank You

I would like to thank A.E.R. Downe who first stimulated my interest in invertebrates and insects in particular and once again thank the past, present, and future students of my Animal Form and Function and Invertebrate Zoology courses here at the University of Ottawa. They have been tremendously patient with my adventures in the use of new technologies in teaching, and after eleven years I, and they, are convinced the adventure was well worth it. I'm would also like to thank reviewers of the *Digital Zoology* program, who provided a thorough commentary that allowed us to make continual improvements to the program and instructor support materials:

Patricia M. Biesiot, The University of Southern Mississippi

C. Leon Harris, State University of New York—Plattsburgh

Also, we extend our gratitude to the many users—students and instructors—of previous versions of the *Digital Zoology* program who provided useful comments to us through the *Digital Zoology* website, and through user surveys. It is our goal to make each version of *Digital Zoology* more expansive and exciting than the versions before, and your suggestions help us reach that goal.

Thanks to Anne Scroggin for her copyediting of the *Digital Zoology* CD and student workbook, and thanks to the folks at McGraw-Hill who have helped to make this all happen: Marge Kemp, Donna Nemmers, Dianne Berning, Linda Avenarius, Renee Russian, and Audrey Reiter, who were tremendously patient as the adventure unfolded. Thank you to the people at N8 Digital, especially Dedee Templeton and Pete Matthews, for their tremendous job in bringing *Digital Zoology* to the web.

As you progress through *Digital Zoology* Version 2.0, please let me know your comments. If there is something I can do better, or something that you would like to see added to the program, let me know. You can get in touch by using the *Digital Zoology* website at www.digitalzoology.com.

— Jon G. Houseman
April 2002

A GUIDE TO USING DIGITAL ZOOLOGY

Digital Zoology Components

When *Digital Zoology* first launches, a dendrogram (tree) of the major animal and protozoan phyla appears on the screen (fig. 1). This is one of four different views of the phyla covered in *Digital Zoology*. They include dendrograms such as this, as well as cladograms based on either traditional morphological data or molecular data. To change between the different views, choose the appropriate button from the choices in the upper-right bar on your screen. Dendrograms show all the phyla in *Digital Zoology*, while the cladogram includes only the animal phyla.

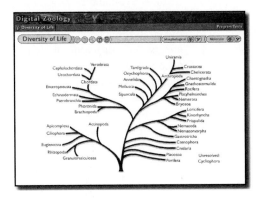

Figure 1: Main dendrogram (tree) for *Digital Zoology*.

Taxon "Read Abouts"

Selecting any of the phyla names listed in the dendrograms or cladograms takes you to the "Read About" for that taxon (fig. 2). Each "Read About" includes information for that taxon with the number of living and extinct species, habitats where they are found, and how they function: locomotion, excretion, osmoregulation, feeding, digestion, respiration, circulatory, nervous, sensory, and reproductive systems are all covered. Highlighted terms are linked to an interactive glossary that includes a definition of the term and information on its derivation. More detailed information on the defining differences unique to the taxon is available by selecting "Defining Differences" from the menu on the left. The same menu includes links to other taxa within the phylum and a direct link to laboratory modules available.

Figure 2: Taxon "Read About".

Laboratory modules

Each laboratory module (fig. 3) opens with an introduction to the lab, providing a brief introduction to what you'll see in the lab, along with some of the unique features available in *Digital Zoology*. Like the "Read Abouts," highlighted terms in the introduction are linked to the interactive glossary. Laboratory modules consist of different topics that may be further subdivided. For example, "Internal anatomy" may include: "General features," "Anterior region," and "Posterior region." You can move to any part of the lab by using the menu on the left-hand side of the screen. Once inside a lab, click on the screen numbers or arrows to move forward or backward one screen at a time.

Figure 3: Laboratory module. Navigate to the different parts of the lab using the menu on the left. Advance through the screens by clicking on the slide number or the left and right arrows.

Interactive cladograms

Cladograms for the phyla at the start of *Digital Zoology*, or classes included in a phylum (fig. 4), provide an opportunity to learn about key events in the evolutionary history of the group. Clicking on a question mark button opens a brief description of the changes that occurred at that point in the evolution of the group. Click on the buttons in order starting from the base of the cladogram and follow the events that defined the group at the end of each branch. Key words in each explanation are linked to the interactive glossary. Clicking on the taxa names at the ends of the cladogram branches will take you to the "Read About" for that group.

Figure 4: Interactive cladogram showing the changes that occurred during the evolutionary history of different groups within the taxon.

Image galleries

For selected phyla, *Digital Zoology* has image galleries (fig. 5) that include photographs and video clips displaying the diversity associated with the phylum. If an image gallery is available for the group you're currently working with, the image gallery button will go from gray to a solid color. Click on the smaller image at the top of the screen to select the image that you want to view. Those with a small film icon in the upper-right corners are video clips. Explanations point out important features about each image or video.

Figure 5: Image gallery.

Self-evaluation

Drag-and-drop quizzes: Laboratory modules have drag-and-drop quizzes that test your familiarity with material in that module (fig. 6). If a quiz is available, you'll see it listed in the navigation bar on the left. The legend for each of the pin heads is given in the list, and below it a set of pin heads with labels is provided. Click and hold the left mouse button down over the labeled pin head, and drag it into the circle that corresponds to the structure identified in the legend. To change your answer, drag the lettered pin head off the circle. To check out your identifications, use the "Submit" button to correct the quiz. Use the right arrow on the question navigation bar to move to the next question. Each time the quiz runs, the questions are slightly different--you might want to try it more than once!

Figure 6: Drag-and-drop quiz module.

Interactive crossword puzzles: Each of the main phyla in *Digital Zoology* has an interactive crossword puzzle (fig. 7) that allows you to test your knowledge of key terms and concepts. The puzzles are built using fill-in-the-blank questions or questions that require only a one- or two- word answer, similar to what you might encounter on an exam. In this format, clues you already have completed may provide letters that help answer another question. Check your answers with the "Check" button that removes any incorrectly placed letters. If you're stuck, reveal a single letter by using the "Reveal letter" button; if you are really stuck, get the answer with the "Reveal" button.

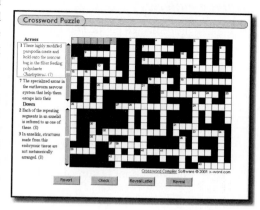

Figure 7: Interactive crossword puzzle.

Additional study suggestions: There are other ways that you can use *Digital Zoology* as a study tool. To test how well you recognize different structures in the lab modules, with a piece of paper cover the legend that appears on the lower-right corner of the screen. Write a legend based on the labels in the picture and then compare your answers to the original. You can also use *Digital Zoology* to test your familiarity with key zoological terms. When you read through any of the explanations and see a term highlighted in blue, write a short definition. Then click on the word and compare your definition to the one that appears on the screen.

Navigation

Digital Zoology runs using an Internet web browser but the usual back and forward buttons of the browser aren't used in navigating through the program. Instead, a series of buttons and menus are used to move between the different parts of the program. A short reminder of what each button does is available by holding the mouse pointer over each button. You can also use Help, available in the Program Tools, to take a short tour that explains how *Digital Zoology* works.

Buttons

 Cladogram button: Click on this button to display the cladogram for the taxon that has been selected. If a phylum's cladogram button is clicked, the cladogram for the classes in that phylum appears. On the main tree, this button switches from the dendrogram view of all the phyla in *Digital Zoology* to the cladogram view of the animal phyla.

Dendrogram button: Clicking on this button on the main tree switches from the cladogram view of the animal phyla in *Digital Zoology* to the dendrogram view of all the phyla.

 Crossword button: Click on this button when it is in color to open an interactive crossword puzzle that uses clues based on the current phylum. When a crossword isn't available, the button appears gray.

 Image gallery button: Click on this button to open an image gallery that includes additional photos and videos that display the diversity found in the current phylum. When an image gallery isn't available, the button appears gray.

Bookmark button: Click on this button to bookmark any of the pages in *Digital Zoology* and use your personal list of bookmarks to jump to any part of the program. This feature is available only online at www.digitalzoology.com.

Notes button: Click on this button to take notes. Enter any reminders or additional information that you may have and then access your set of notes along with where in *Digital Zoology* the notes were taken. This feature is available only online at www.digitalzoology.com.

Question button: Click this button on a cladogram for an explanation of the changes that occured at this point in the evolutionary history of the taxon.

Explanation button: You can't click on this button on a cladogram, but it will tell you the point on the cladogram that corresponds to the explanation that appears on the screen.

Menus

A green taxon bar runs across the top of the screen underneath the blue title of *Digital Zoology*. It starts on the far left with Diversity of Life, and as you move through the different phyla the taxon bar expands to identify the kingdom and infrakingdom for the phylum you are looking at. Select any of the taxonomic names in the taxon bar to navigate to that group of organisms. A second menu appears on the left side of the screen. Its content varies depending where you are in the program, and if you click on any of the items listed there you'll move to that part of the *Digital Zoology*.

Program tools

The Program Tools menu is located at the far right of the green taxon bar. Program Tools include another link to the **Bookmark** and **Notes** tools available in the online version of *Digital Zoology*. **Glossary** in the menu opens the full glossary, allowing you to browse and search for specific terms. By selecting **Bibliography** you can access a list of key resources used to create *Digital Zoology*; **Credits** lists the people who put *Digital Zoology* together, and provides copyright information for borrowed materials. **Help** gives you a guided tour of *Digital Zoology* and how it works.

DIGITAL ZOOLOGY VERSION 2.0 CORRELATION CHART

	Digital Zoology*	Hickman et al., Integrated Principles of Zoology, 11th ed.	Hickman et al., Animal Diversity, 3rd ed.	Miller & Harley, Zoology, 5th ed. ‡web sites	Hickman & Kats, Laboratory Studies to Accompany Integrated Principles of Zoology, 10th ed.	Lytle, General Zoology Laboratory Guide, 13th ed.	Pechenik, Biology of the Invertebrates, 4th ed.
Protozoa:							
Actinopoda	B,P,V	11	5	8	6	5	3
Ciliophora	B,P,V	11	5	8	6	5	3
Apicomplexa	B,P	11	5	8	6	5	3
Euglenozoa	B,P,V	11	5	8	6	5	3
Rhizopoda	B,P,V	11	5	8	6	5	3
Granuloreticulosa	B,P	11	5	8	6	5	3
Porifera	B,P,Q	12	6	9	7	6	4
Placozoa	B	12	—	9‡	—	—	4
Cnidaria	B,C,P,Q,V	13	7	9	8	7	6
Ctenophora	B	13	7	9	—	—	7
Platyhelminthes	B,C,P,Q,V	14	8	10	9	9	8
Gastrotricha	B,P,V	15	9	10	10	10	18
Gnathostomulida	B	14	8	—	9	—	18
Cycliophora	B			7	—	—	18
Rotifera & Acanthocephala	B,P,Q,V	15	9	11	10	10	11
Annelida	B,C,P,Q,V	17	11	13	12	12	13
Mollusca	B,C,P,Q	16	10	12	11	11	12
Sipuncula	B	21	13	—	—	—	13
Nemertea	B	14	8	10	—	—	10
Bryozoa (Ectoprocta)	B,P,Q	22	13	16‡	—	—	19
Brachiopoda	B,P,Q	22	13	16‡	—	—	19
Phoronida	B	22	13	16‡	—	—	19
Nematoda	B,P,Q,V	15	9	11	10	10	16
Nematomorpha	B	15	9	11	10	10	17
Onycophora	B	21	13	—	—	13	15
Tardigrada	B	21	13	—	—	—	15
Arthropoda:	B,C,P	18	12	14	13	13	14
Chelicerata	B,C,P	18	12	14	13	13	14
Crustacea	B,C,P,Q	19	12	14	14	13	14
Uniramia	B,C,P,Q,V	20	12	15	15	13	14

	Digital Zoology*	Hickman et al., *Integrated Principles of Zoology*, 11th ed.	Hickman et al., *Animal Diversity*, 3rd ed.	Miller & Harley, *Zoology*, 5th ed. ‡web sites	Hickman & Kats, *Laboratory Studies to Accompany Integrated Principles of Zoology*, 10th ed.	Lytle, *General Zoology Laboratory Guide*, 13th ed.	Pechenik, *Biology of the Invertebrates*, 4th ed.
Priapulida	B	15	9	11	—	10	17
Kinorhyncha	B	15	9	11	—	10	17
Loricifera	B	15	9	11	—	10	17
Echinodermata	B,C,P,Q,V	23	14	16	16	14	20
Chaetognatha	B	24	13			—	18
Hemichordata	B,C,P	24	14	17	—	—	21
Chordata:							
Cephalochordata	B,C,P	25	15	17	17	15	22
Urochordata	B,C,P	25	15	17	17	15	22
Vertebrata:	B,C,P	26	16	18	18	—	—
Agnatha	B,P	26	16	18	18	—	—
Chondrichthyes	B,P,Q,V	26	16	18	18	16	—
Osteichthyes	B,P,Q,V	26	16	18	18	17	—
Amphibia	B,P,Q,V	27	17	19	19	18	—
Reptilia	B	28	18	20	20	—	—
Aves	B	29	19	21	21	—	—
Mammalia	B,P,Q	30	20	22	22	19,20	—
* B=basic text, C=cladogram, P=pictures of lab materials, Q=quiz, V=video							

PROTOZOANS

Inside *Digital Zoology*

As you explore the Protozoans on the *Digital Zoology* CD-ROM or on the web at www.digitalzoology.com, don't miss these learning tools:

○ Photos of prepared slides from a wide variety of protozoans including: foraminiferans; rhizopods (*Amoeba, Difflugia,*); actinopods (*Actinosphaerium*); ciliates (*Paramecium, Blepharisma, Stentor, Spirostomum,* Didinium, *Vorticella,* and *Euplotes,* among others); Apicomplexans (*Plasmodium,* the causative organism of malaria); flagellated euglenozoids (*Euglena* and *Trypanosoma*); and a variety of related protests including colonial Volvox.

○ Video clips include: locomotion using flagella, cilia, and pseudopods; feeding; how the water expulsion vesicle works; metachronal waves in ciliates; and a variety of behaviors, including escape reactions involving contraction of the myoneme.

Defining Differences

Some of the differences described in the following sections appear for the first time in the protozoans, and they define the group. Others are important for understanding how these organisms function. Whichever is the case, you'll want to watch for examples of each in the various protozoans that you will be examining.

○ Unicellular "animals" with
○ specialized organelles including
○ undulipodia used for locomotion.

Unicellular "animals"

The single-cell organisms, protists, have always been a problem for taxonomists, who have trouble defining categories for the organisms. Their mixed feeding habits, which include being capable of **photosynthesis - autotrophic**, ingestion of other organisms - **heterotrophic**, or survival on dead and decaying materials - **saprophytic**, makes them single-cell versions of the plant, animal and fungal kingdoms. The kingdom Protista probably shouldn't be a kingdom on its own. Instead, single-celled **eukaryotes** may be better placed in the other biological kingdoms. In a zoology course the protists that we are most interested in have animal-like traits: no cell wall, heterotrophic, and capable of movement - the protozoans.

Protozoans are microscopic in size, although some are large enough to be seen with the naked eye. Microscopic organisms such as these have a number of advantages, one of which is that they have a high surface-to-volume ratio. That's an advantage because diffusion is an important process for many of their functions. Don't let their small size fool you; these are complex organisms. Everything they do, and how they do it, is packed into a single cell. In animals, organs and tissues become specialized for certain functions; for protozoans, it's their **organelles** that take on the same functions.

Like all animal cells, protozoans are covered by a **plasma membrane** that surrounds the cytoplasm of the cell, the protozoan's integument or skin. Like all membranes, it's permeable; water and low molecular weight substances dissolve in it. Oxygen, required for metabolism, and metabolic wastes generated by metabolism, diffuse across the surface, giving the plasma membrane respiratory and excretory functions. The **nucleus**, with its **nucleolus** and chromosomes, contains the genetic material and will be involved in providing copies of itself for new cells, the reproductive system. The **Golgi apparatus, rough endoplasmic reticulum**, and the **lysosomes** that they produce, fuse with ingested **food vacuoles**,

---

phagosomes. In some protozoans, food is always ingested at the same place on the cell surface, the **cytostome** (cell mouth). Undigested food is removed at a **cytoproct**, or the cell anus. All of this is similar to a digestive system. **Mitochondria**, which fuel life's processes in all cells, use the nutrients from digested food and oxygen that has diffused across the cell surface. In freshwater protozoans the **water expulsion vesicle** is an osmoregulatory organelle removing excess water that diffuses into the cell. Protozoans have **endoskeletons** formed from the microtubular cytoskeleton of the cytoplasm or exoskeletons secreted as shells or **tests**. Some of these single-cell organisms are capable of defending themselves using organelles such as **trichocysts**, and others can orient to external stimuli and detect light with **stigmata**. Movement is even controlled by organelles, **cilia** and **flagella**. These and the **pseudopodia** in the amoebas give these small creatures movement, something we associate with animals. Essentially, everything that a large animal can do, single-cell protozoans can do as well, only on a much smaller scale.

Specialized organelles

$\mathbf{Y}$ou won't be able to see all these different organelles with the light microscope, so we'll focus on those you'll probably see in your lab or *Digital Zoology*.

The nucleus contains the genetic information of the cell. In addition to activities that maintain the cell, the nucleus creates copies of itself for new protozoans by either **asexual reproduction** - using **mitosis**, or **sexual reproduction** - using **meiosis**. Asexual reproduction often involves **fission**, or some sort of division of the parent cell into daughter cells. If the two daughters are similar in size, it's **binary fission**. If one's large than the other, it's **budding**. If there has been multiple mitotic divisions and the parent cell divides into more than two daughters, it's **multiple fission**. Some protozoans form gametes that fuse to form a new protozoan. Ciliates have two nuclei: a **macronucleus** for running the day-to-day functions of the cell and a **micronucleus** for reproduction. Ciliates have a form of genetic mixing, **conjugation**, not found in the other protozoans.

✍ *What are the similarities and differences between sexual reproduction and conjugation?*

The water expulsion vesicle (also called a contractile vesicle) is an important osmoregulatory organelle in freshwater protists, which are **hyperosmotic** to the surrounding water. Water constantly diffuses into the cell and could cause the cell to burst if there wasn't some way to get it out. That is what the water expulsion vesicle does. A network of collecting tubules extends throughout the cells, and the rich supply of mitochondria on the surface of the tubules provides the energy needed to pump water from the cytoplasm into the tubules. The tubules dump the water they've collected into the water expulsion vesicle, which slowly fills. Once full, the vesicle's membrane fuses with the outer plasma membrane creating a pore out of which the excess water flows.

Protozoans ingest food by **phagocytosis**, and as a result, the ingested food is surrounded by a plasma membrane creating a food vacuole, or phagosome. Digestion doesn't start until the phagosome fuses with a lysosome that contains the digestive enzymes. Everything being digested, and the enzymes digesting it, is contained inside a plasma membrane permeable only to small molecular weight compounds, the products of digestion. The enzymes are too big to cross the membrane, so there's no danger that they might mistakenly digest the cytoplasm of the cell. Initially the pH inside the food vacuole is acidic and as

---

digestion proceeds, the contents of the vacuole become more alkaline. To be sure that nutrients get to every part of the cell, the food vacuole is swept around the cell by the cytoplasmic currents and nutrients diffuse out supplying the different parts of the cell. Once there is only undigested food left, the food vacuole fuses with the cell membrane, dumping its contents to the outside of the cell.

Undulipodia: Cilia, flagella, and pseudopods

Cilia and flagella share a common structure, with a microtubular core that has a 9+2 organization with nine paired doublets of **microtubules** surrounding a central pair to form the **axoneme**. The axoneme is surrounded by plasma membrane and anchored to the inside of the cell by the **kinetosome** formed from nine microtubular triplets connected to the nine doublets in the axoneme. Inside the axoneme arms, the nine doublets extend arms to their neighbors, and when ATP is burned, the position of these arms changes and the microtubules slide past each other. That sliding motion is what bends the axoneme and in turn, the cilium or flagellum bends.

✍ *What are the differences between cilia and flagella?*

In protozoans flagella are always found one per cell. Cilia are much more numerous and, depending on where on the cell surface they are found, they can be modified for different functions. Cilia can fuse to make even bigger structures, creating **complex ciliature** such as **cirri** and **membranelles**.

Cilia and flagella can generate substantial force as they try to pull or push a protozoan through the water. To be able to do that, the plasma membrane needs to be strengthened, and when it is, the surface is referred to as the **pellicle**. In some ciliates even further strengthening occurs with the kinetosomes being anchored to each other creating a complex infraciliature. The stronger the pellicle, the less likely it is that the shape of the protist can change, and most ciliates have recognizable shapes. A good example is *Paramecium* with its oval shape and food groove.

Movement in a pseudopod involves changes in the fluidity of the cytoplasm from the more solid **plasmagel (ectoplasm)** or fluid **plasmasol (endoplasm)**. Like muscles in higher animals, pseudopodial movement involves actin, myosin, and ATP.

✍ *Describe the biochemical events involved in changing endoplasm to ectoplasm and back to endoplasm.*

Kingdom Protista

The classification of the kingdom Protista seems to be constantly changing, and no sooner is one classification scheme published then another appears with a revisions or changes. There is a growing consensus that mitochondria may be the clue to some stability in the major taxa, at least at the higher levels such as infrakingdom. The first eukaryotes enclosed their genetic material in a plasma membrane, and acquisition of the **prokaryote symbiont** that ultimately became the mitochondria probably happened not long after that. Mitochondria generate the energy to drive life's processes by the **electron transport chain** embedded in the mitochondrial **cristae**. The most primitive protists are amitochondrial (subkingdom Archezoa), and those with mitochondria (subkingdom Neozoa) have three different types of mitochondrial cristae: tubular, flattened, or disc-shaped.

Infrakingdom Sarcomastigota

Most of the protists in this taxon are free-living heterotrophs that move using pseudopodia for amoeboid movement and feeding. Their pseudopods include a variety of different types: **lobopodia**, **filipodia**, and **reticulopodia**, and unlike the other protozoans, sarcomastigotes are variable in shape. Sarcomastigotes reproduce primarily by either binary fission or multiple fission, and the cristae in their mitochondria are tubular, sometimes flattened, but never discoid.

Infrakingdom Alveolata

The alveolates include of three groups: ciliates, heterotrophs; apicomplexans, parasites; and dinoflagellates, autotrophs. Only the first two qualify as protozoans. Despite their varied lifestyles, the alveolates all have tubular cristae in their mitochondria and a series of complex folded membranes that form pockets, alveola, under the outer plasmalemma. Alveolates also have extrusomes, trichocysts for example, used to capture prey or to defend against predation.

Infrakingdom Euglenozoa (Infrakingdom Discicristata or Phylum Discomitochondria)

As you can see there isn't agreement on the classification of this protist taxon. Some define the group of organisms with their flattened disc-shaped cristae in their mitochondria as an infrakingdom with two phyla, each with two classes, and some define the group as a phylum of four classes. In either case, at class level the groups are similar in both taxonomic schemes and include protists formerly described as the flagellates or zoomastigophorans. If we use the phylum designation, then the four classes are the Amoebomastigota, Kinetoplastida, Euglenida, and the Pseudociliata. The kinetoplastids and euglenids are commonly studied in zoology courses and are closely related to each other as the phylum Euglenozoa in the infrakingdom Discomitochondria.

Infrakingdom Atcinopoda

Depending on the classification scheme, these radially symmetric protists are either an infrakingdom with two phyla, Heliozoa and Radioza, or a phylum divided into four classes: Heliozoa; Pheodaria and Polycystina, which together are the radiolarians; and marine Acantharia, which may also be radiolarians - part of the reason for the phylum Radiozoa.

A Closer Look Inside *Digital Zoology* - Protozoans

Prepared slides and wet mounts of live specimens are the best way to examine the protozoans. Internal structures are easier to see in prepared slides because histological stains highlight nuclear and proteinaceous structures. Prepared slides don't, however, give a sense for how the organisms move - that's the value of the wet mount. If you can, try to look at the two preparations together. Watch your specimen move, then look for the internal structures. If live and prepared specimens aren't available, you'll find the combination in the photos and videos in *Digital Zoology*.

Granuloreticulosea, the foraminifera

These protozoans are commonly referred to as forams, and they, like the actinopods, are best known for their shells, only in this case they're usually made of calcium salts. A foram's shell is multichambered, and each chamber is closed with successive chambers arranged in a spiral with each a bit larger than the one preceding. You can see examples of the shells in *Digital Zoology*. The cell body is located inside the chambers and the netlike reticulopodia extend through the small openings in the walls of the shell and are involved in feeding and locomotion. Almost all these protozoans are marine, and their shells cover the ocean bottoms in shifting ooze, or strew. You might be more familiar with the fossil version of this ooze, chalk. The chalk cliffs of England and the stone of the great pyramids of Egypt are examples of fossilized foramaniferans.

Rhizopoda, the amoebas

Rhizopods are amoebas, and there are two types: naked and shelled. It is the naked amoebas with which you might already be familiar.

As you examine the prepared slides you should be able to see the water expulsion vesicle, nucleus, food vacuoles, and pseudopods. In a wet mount some of these organelles will be visible, but with this preparation watch the pseudopods move. At the tip is the clear zone, the **hyaline cap**, which the fluid endoplasm pushes against before flowing to the edge of the pseudopod and turning into ectoplasm. Look at the other end of the amoeba and you'll see the reverse, ectoplasm turning back to endoplasm then flowing into the center of the advancing pseudopod. If you don't get a good view of it in your lab, check in *Digital Zoology*.

✍ *How many different types of pseudopods can protozoans have?*

The tests of shelled amoebas are made from a variety of substances including chitin, sand grains, and substrate. Be sure to take a look at *Difflugia*, an example of shelled amoebas.

Actinopoda

Actinopods are easily identified by their unique pseudopods, axopods, with a central, supporting axoneme of microtubules. Axopoda extend in all directions from a central test, and each has a thin layer of cytoplasm streaming along its length. It's not a surprise that freshwater species are called sun animalicules. Take a look at the photos and video *Actinosphaerium* in *Digital Zoology* to see the delicate axopods, which in prepared slides are often bent and damaged. The central test in some marine forms is made of silica, and the discarded shells that accumulate in the ocean depths are referred to as strews. Radiolarians are a taxon in the phylum, and in *Digital Zoology* you can see a sample of the bewildering array of designs that these protozoan shells can have.

✍ *How does an actinopod feed?*

Ciliophora, the ciliates

One of the most often used ciliates in zoology courses is *Paramecium*, and no doubt you'll get to see it and a few others in your lab session. *Digital Zoology* includes *Paramecium*, *Blepharisma*, *Stentor*, *Spirostomum*, *Didinium*, *Vorticella*, *Euplotes*, along with a quick look at a few others.

We're looking at ciliates, so one of the things that you want to be sure to see are the cilia that cover the cell. Cilia and their kinetosomes are anchored into the plasma membrane with a reinforcing pellicle, and it's the reason why many ciliophorans have a consistent shape, or look. You can see it on the specimen as lines that appear to run the length of the organism. *Blepharisma* and *Spirostomum* are good examples for seeing the pellicle.

During phagocytosis the cell membrane surround the food without disrupting the reinforcing pellicle. The solution is to always ingest food at the same place on the cell surface, the cytostome. The other side of the coin is that undigested food is also removed from the ciliate at a cytoproct. In the simplest ciliate, the ciliature is the same all over the body; in others it becomes specialized. With feeding always happening at the same place on the cell surface, the cilia in that area become modified for capturing food and are called oral ciliature. Somatic cilia cover the rest of the body and are involved in locomotion. Good examples of modified cilia and the cytostome can be found in *Paramecium*, *Blepharisma*, *Stentor*, and *Vorticella*. Be sure to take a look at the formation of the food vacuole in *Paramecium* available on the *Digital Zoology* CD-ROM.

Cilia on the cell surface move in a coordinated manner. Not all of the cilia are at the same point in the power and recovery stroke, a **metachronal wave** moving across the surface as the cilia beat. Although you can see it on a number of the specimens in *Digital Zoology*, the wave that runs around the lip of *Stentor* is a good example. While some swim using cilia, other ciliophorans remain attached to the substrate - *Stentor* and *Vorticella*, for example. Both specimens have contractile elements and are able to move using these; watch *Vorticella* with the coiled stalk and its internal **myoneme**.

The macronucleus is easy to see in the prepared slides and the living specimens. If you don't get to see a water expulsion vesicle empty during the lab, be sure to see it with *Paramecium* and *Vorticella* in *Digital Zoology*.

Apicomplexa

The causative agent of malaria, *Plasmodium*, is an example of an apicomplexan. Like all parasites, what a disease organism is, this protozoan has a complex life cycle. This complexity is, in part, so the **parasite** can move from one host to another. (Parasitism is discussed in more detail in the platyhelminth chapter.) Malaria moves from mosquitoes to humans and back. If the different stages of the life cycle aren't available, you'll find pictures of **sporozoites**, **merozoites**, **trophozoites** (inside the red blood cells), and oocysts in the gut wall of the mosquito in *Digital Zoology*.

At one time it was thought that the world was free of malaria because the mosquito that carried it could be controlled with pesticides, which at that time included DDT. Unfortunately the announcement was premature. The mosquito survived and is now resistant to insecticide treatments. From only a few reported cases a year in the 1960s, malaria now kills more than three million people annually.

 What are the different stages in the malaria's life cycle, and what is the functional purpose of each?

Euglenozoa

Euglena is often used as an example of a protozoan that swims using a flagellum. If you set up your microscope properly, with not too much light shining on the specimen, you'll be able to see the flagella and stigmata, or eye spot, which the species uses to orient toward light. *Euglena* sits on the fence between being a unicellular animal and/or a plant. It is capable of photosynthesis, that's why it's green; but if there is no light, it can survive as a heterotroph. Like ciliates, *Euglena* has a pellicle to anchor the flagellum to the cell. *Euglena's* pellicle isn't as rigid as a ciliate's; that allows for the euglenoid movement that you may get a chance to see. If not, there's a video of it in *Digital Zoology*. Another flagellate example is *Trypanosoma*, which causes sleeping sickness. The trypanosome is found between the red blood cells. The base of its flagellum is anchored at the anterior end and is fused along the length of the cell in an undulating membrane that looks almost like a fin.

✍ *Describe the life cycle of the causative organism of sleeping sickness,* Trypanosoma cruzi.

Related protists

Volvox is an example of a colonial phytomastigophoran, and it's often included in zoology courses because some believe it gives clues to how multicellular metazoa first appeared. Many metazoans go through a hollow ball stage, the blastula, during their embryology. The assumption is that colonial *Volvox* may be some sort of prototype for that stage. Spherical and radial symmetries are often also thought to be primitive traits in the ancestry of the metazoa. If that's the case, again *Volvox* may provide another clue to metazoan origins. (Symmetry is discussed in more detail in the cnidarian chapter.) There are problems with this, not the least of which is that *Volvox* is a colony of diflagellate, photosynthetic algae, with haploid somatic cells. These traits are never seen in animals!

Cross-References to Other McGraw-Hill Zoology Titles

Integrated Principles of Zoology, 11th edition. C. P. Hickman, L. S. Roberts, and A. Larson. Chapter 11.

Animal Diversity, 3rd edition. C. P. Hickman, L. S. Roberts, and A. Larson. Chapter 5.

Zoology, 5th edition. S. A. Miller & J. P. Harley. Chapter 8.

Biology of the Invertebrates, 4th edition. J. Pechenik. Chapter 3.

Laboratory Studies in Integrated Principles of Zoology, 10th edition. C. P. Hickman, F. Hickman, and L. Kats. Chapter 6.

General Zoology Laboratory Guide, 13th edition. C. Lytle. Chapter 5.

Structures Checklist

Here are some of the structures that you should be able to easily find in *Digital Zoology* and the specimens that you will be looking at in your lab. After reading your lab handout, you might want to add more and, depending on the equipment available in your lab, you might see more. As you study the material, you might also want to make some notes on how some of these structures looked or include a drawing in your lab notes. (Structures indicated by * may be hard to see.)

Granuloreticulosa

- ☐ Foraminiferan strew

Rhizopoda - slides and/or wet mounts

Amoeba (slide and wet mount)

- ☐ Ectoplasm
- ☐ Endoplasm
- ☐ Food vacuole
- ☐ Hyaline cap

- ☐ Nucleus
- ☐ Plasmalemma
- ☐ Pseudopod (lobopod)
- ☐ Water expulsion vesicle

- ☐
- ☐

Difflugia

- ☐ Pseudopod
- ☐ Pylome

- ☐
- ☐

- ☐ Test (shell)

Actinospaerum

- ☐

Actinopoda - slides and/or wet mounts

Actinospaerum

- ☐ Axopodia
- ☐ Cell body
- ☐ Ectoplasm

- ☐ Endoplasm
- ☐ Water expulsion vesicle
- ☐

- ☐
- ☐
- ☐
- ☐

Additional structures

- ☐ Radolarian strew
- ☐
- ☐

Ciliates - slides and/or wet mounts

Paramecium

- ☐ Binary fission
- ☐ Cilia
- ☐ Conjugation
- ☐ Cytostome
- ☐ Food vacuole
- ☐ Macronucleus
- ☐ Micronucleus*
- ☐ Oral groove
- ☐ Pellicle
- ☐ Water expulsion vesicle
- ☐
- ☐
- ☐
- ☐
- ☐

Blepharisma

- ☐ Cilia
- ☐ Complex ciliature
- ☐ Macronucleus
- ☐ Oral groove
- ☐ Pellicle
- ☐ Water expulsion vesicle
- ☐
- ☐

Stentor

- ☐ Basal attachment
- ☐ Cilia
- ☐ Complex ciliature
- ☐ Cytostome
- ☐ Food vacuole

- ☐ Gullet
- ☐ Macronucleus
- ☐ Pellicle
- ☐ Stalk
- ☐ Water expulsion vesicle
- ☐
- ☐
- ☐

Vorticella

- ☐ Cilia
- ☐ Complex ciliature
- ☐ Cytostome
- ☐ Food vacuole
- ☐ Macronucleus
- ☐ Pellicle
- ☐ Stalk

- ☐ Water expulsion vesicle
- ☐
- ☐

Additional observations

- ☐ Metachronal waves
- ☐ Cytoplasmic streaming
- ☐
- ☐
- ☐
- ☐
- ☐

Apicomplexa

Malaria -Prepared slides

- ☐ Merozoites
- ☐ Oocysts

- ☐ Red blood cells (host)

- ☐ Sporozoites
- ☐ Trophozoites

☐   ☐
☐   ☐

Euglenozoa - slides and/or wet mounts

Euglena

☐ Cell body
☐ Chloroplast
☐ Contractile vacuole*
☐ Flagellum*

☐ Nucleus*
☐ Paramylon body
☐ Pellicle
☐ Plasmalemma
☐ Stigma
☐

☐

Trypanosoma

☐ Flagellum
☐ Nucleus
☐ Plasmalemma

☐ Red blood cells (host)
☐ Undulating membrane
☐

Related protists

Volvox

☐ Asexual colony
☐ Daughter colony
☐ Parent colony

☐ Protoplasmic strands
☐ Reproductive cells*
☐ Sexual colony
☐ Somatic cells

☐
☐
☐
☐

Crossword Puzzle – Protozoans

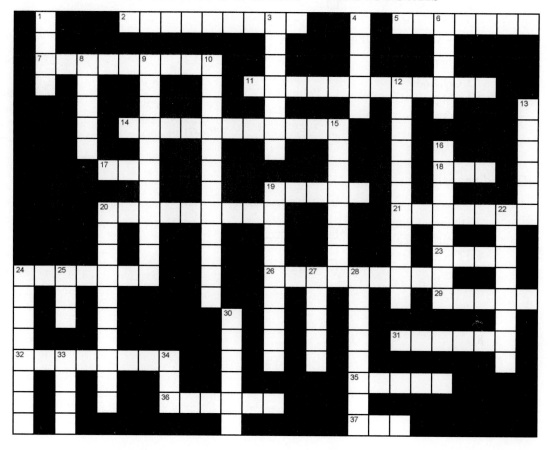

Puzzle solution. An interactive web-based version of this puzzle and its solution are available on the *Digital Zoology* website www.digitalzoology.com/students. With the interactive puzzle you can check to see if individual words or the whole puzzle is correct and get hints for single letters.

Across

2 This is the more fluid of the two types of protoplasm found in an amoeba. (9)

5 Trichocysts are used for this in ciliates. (7)

7 Food always enters a ciliate at the same point on its surface, referred to as this. (9)

11 This nucleus is the larger of the two different nuclei found in ciliates. (12)

14 This stage of the malarial life cycle is found in the salivary glands of the mosquito host. (11)

17 The number of conjugants involved in conjugation in ciliates. (3)

18 Although ancestrally they may have had more, almost all flagellate protozoans have this number of flagella. (3)

19 Membranelles of an oligohymenorans are made up of these. (5)

20 In apicomplexan protists, sporozoites result from this reproductive process. (9)

21 This supports an amoeba's axopod. (7)

23 This type of ciliature around the cytostome is important for identifying the different ciliophoran taxa. (4)

24 This cap is found at the advancing edge of a pseudopod. (7)

26 With its multiple copies of the genome, this describes the macronucleus in a ciliate. (9)

29 Amoebas consume this type of food by using phagocytosis. (5)

31 One of the two ways that flagella beat. (6)

32 Single cilia often fuse to form larger structures collectively called this type of ciliature. (8)

35 Complex ciliature involved in walking in some ciliates. (5)

36 Apicomplexan protists have this type of complex that defines the group. (6)

37 The larger of the two products of gametogony. (3)

Down

1 The part of the amoeba where the ectoplasm is converted back into endopalsm. (4)

3 These cilia on a ciliate's body are not involved in feeding and are referred to as this type of ciliature. (7)

4 The part of the amoeba where the endoplasm is converted to ectoplasm. (5)

6 The sudden release of the malarial parasite causes this in a patient suffering from malaria. (5)

8 Not all amoebas are naked, some live inside these. (5)

9 This stage of the malarial parasite is found in the red blood cells. (11)

10 In ciliates, once the nuclear material has been exchanged, the conjugants separate and are now called these. (12)

12 The process of sexual reproduction in ciliates. (11)

13 The pseudopods extend through this opening in the amoeba's test. (6)

15 The malaria parasite lives in these mosquito glands and are injected into the human host when the mosquito bites. (8)

16 The broad, fleshy pseudopods that we usually associate with amoebas are of this type. (8)

19 Undigested food exits from the same point on a ciliate; it's referred to as this. (9)

20 Describes the cycle of multiple fissions by the malarial parasite inside the human host. (10)

22 Schizogony is a form of this type of fission. (8)

24 *Euglena* uses this type of flagellar beat to pull its way through the medium. (8)

25 Like the muscles in higher animals, amoebas also depend on contractile proteins, calcium, and this for movement. (3)

27 This is the first type of cell is attacked by the malaria parasite as it enter its human host. (5)

28 The kinetodesmata wind together in ciliates to form this. (8)

30 In the deepest ocean depths, the shells of amoebas can only be made of this. (6)

33 In reticulopod amoebas, their pseudopods form this. (4)

34 Conjugation is a unique way of mixing this from two ciliates. (3)

PORIFERA

Inside *Digital Zoology*

As you explore the Porifera on the *Digital Zoology* CD-ROM or on the web at www.digitalzoology.com, don't miss these learning tools:

- ❍ Photos of cross sections and dissected specimens of preserved sponges *Grantia (Scypha)* and *Leucosolenia* and a variety of other dried sponges show the diverse shapes that sponges have and how their spicules are organized.
- ❍ Drag-and-drop quiz on the key structures of the sponge *Grantia (Scypha)*.

Defining Differences

Some of the differences described in the following sections appear for the first time in the Porifera, and they define the phylum. Others are important for how animals in the phylum function. Whichever is the case, you'll want to watch for examples of these in the various sponge specimens that you will be examining.

- ❍ Cellular grade of organization with
- ❍ totipotent cells, including
- ❍ choanocytes in an
- ❍ aquiferous system.

Cellular grade of organization

That a sponge is an animal may come as a surprise you. You're not alone. Biologists considered them plants until the late 1700s. Sponges are multicellular animals but with a difference. That difference is why they are referred to as **parazoans,** rather than **metazoans**. They are "almost animals" because they are missing one thing that all other metazoans have–cells organized into tissues, and this is why they are described as being functionally organized at the **cellular grade**.

Sponges are an example of one of two ways that animals became multicellular–by using either cells organized into tissues or as aggregates of independent cells. Sponges are an example of the second, and although it's a body plan that serves them well for what they have to do, its dependence on independent cells limits their potential for diversification. It's one of the reasons why you'll see sponges referred to as an evolutionary dead end.

Don't become confused, though, when you are looking at the sponge cross sections in *Digital Zoology* or your lab. You'll see an outer layer of **pinacoderm** cells and a layer of **choanoderm** cells lining the **spongocoel.** These are separated from each other by the **mesohyl**–but remember these are only layers, not tissues. The reason that they aren't tissue is that cells in a true tissue layer sit on a **basement membrane** and have a variety of cell junctions for intercellular communication–both of these are missing in the sponge cell layers.

Totipotent cells

Even though a sponge's cells are only arranged in layers, there is still a division of labor between them, with certain cells responsible for one function while others have different functions. What all the cells have in common is that sponge cells are **totipotent**, meaning each of the cell types can, if needed, differentiate into the basic **archaeocyte** cell and then redifferentiate into a whole new type of cell. It's part of how sponges respond to changes in their environment and the reason for their often asymmetric shapes. They can't move, so they grow toward where they want to be. If they're damaged they can grow back what was lost. This is an effective strategy, but its dependence on totipotency also prevents the development of tissues, a characteristic of all the other multicellular body plans that follow.

 What are the different types of cells that you would find in a sponge, and what does each cell type do?

Choanocytes

Pumping water and filtering food from it is what sponges do best, and it's their **choanocytes** that take care of that. These specialized cells have a central flagellum surrounded by a **collar** of **microvilli** connected to each other by an even finer mesh of **microtubules**. As the central flagella beats, water is drawn in toward the collar and between the microvilli and its **microtubular mesh**. Only water can pass through, and small particles of food carried in it become trapped on this delicate sieve. The choanocytes then use phagocytosis to consume the trapped food.

It seems so simple, but you multiply this by the huge numbers of choanocytes in a sponge and you have lots of water being pumped and just as much food being filtered from the water. Sponges were some of the first filter feeders in the ancient oceans, and the reason sponges have survived is that, in many ways, they still are the best at filter feeding.

How does captured food get from a choanocyte to the other sponge cells?

Aquiferous system

There's more to it than pumping and filtering water, however. The nutrients need to be extracted from the water. That's the role of the **aquiferous system**, which combines pumping and food capturing choanocytes with a variety of canals, chambers, and spaces through which water flows. In sponges the aquiferous system is arranged into three different architectures: **asconoid**, **syconoid**, and **leuconoid**.

In the simplest, the asconoid, choanocytes line the spongocoel, and water is drawn in through cells in the body wall. There's always a possibility that some water could make it to the central spongocoel without being filtered. In the syconoid architecture, the filtering choanocytes are found in smaller radial canals that empty into the central spongocoel, which is a better way to capture particulate food.

But as the diameter of the passages through which the water flows narrows, and the same volume is pumped, the water speeds up. It's like when you put your finger over the end of the garden hose to squirt the water farther. It works because the water flows faster through the narrower opening at the end of the hose. Syconoid sponges are better at trapping and pumping than the asconoid architecture but, there's still room for improvement.

That improvement is the leuconoid sponge. Here the narrow tubes through which the water flows widen into chambers where the choanocytes are located. Water rushes through the sponge and slows in the chambers where it is filtered. It leaves the chamber and travels through more of these choanocyte spaces before leaving the sponge through the multiple oscula. Now you can see why most of the sponges have the leuconoid architecture and why we said earlier that sponges are pretty good at what they do.

 Use the following terms and place them in the proper order to explain the flow of water through the three different sponge architectures: spongocoel, osculum, radial canal, excurrent canal, prosopyle, apopyle, choanocyte chambers, porocyte.

A Closer Look Inside *Digital Zoology* - Porifera

Leucosolenia

Don't let the name fool you. *Leucosolenia* is not a leuconoid sponge! It's an example of the simplest of sponge architectures. The vase shape may be hard to see in most preserved specimens because the species is colonial. The central spongocoel of each sponge is connected to the others by hollow horizontal branches, extensions of the spongocoel. Either with the specimens you have in the lab or the photos in *Digital Zoology* take a look at the specimen, and you'll see that it's made of delicate spicules over which is woven a fine film of living tissue, making it difficult to identify the different cell types.

Grantia *or* Scypha

Either *Grantia* or *Scypha* are excellent examples of this intermediate architecture and a chance to see the cell layers of a sponge. What most students find hard when observing this specimen is that it is a modified version of the syconoid architecture. Look closely at the surface of a preserved specimen and you'll see that it's covered in fingerlike projections. The hollow inside of each finger is the radial canal and the spaces between the fingers are the incurrent canal. In the dissected specimen in *Digital Zoology* you can see the inner surface of the spongocoel and the openings of the radial canals contained in the fingers that lead to the spongocoel.

Another problem in understanding how this specimen works is apparent when looking at the cross section. As microscopic sections are made, the cut isn't perfect, leaving a nice petal-like array of radial canals surrounding the central spongocoel. Instead, you'll have to look around on your specimen to see the connection between the spongocoel and the radial canal. It is always a good idea to look at more than one section whenever possible. Once you can identify the inside of the radial canal, you'll see the thicker choanocytes on the inner surface. If you have good samples and an adequate microscope, the flagella may be visible, but don't expect them to stick out like they do in the drawings. Instead they usually appear as a tangled mat, at or near the cell body of the choanocyte.

A second problem with the sectioned materials is that instead of cutting through the spicules, the blade often breaks them, and this tears the adjacent tissue. Move along the inner surface of the radial canal to find the connection between it and the incurrent canals. Be patient; it's there!

Other sponges, spicules, and gemmules

Almost all living sponges have this leuconoid architecture, but it's difficult, especially with dried specimens, to see all the elements of their design. Larger canals that lead to the osculum are much easier to identify than the flagellated chambers.

Sponges support themselves with spicules, and the different classes in the phylum are defined by their spicules. You'll have seen broken spicules in *Grantia* or *Scypha* cross sections. Take a closer look at the surface of some of the other sponges or microscope slides of prepared spicules to see some of the different shapes. Perhaps the most elegant spicules are the six-sided spicules that glass sponges weave into their elaborate skeletons. You'll find some good examples on the *Digital Zoology* CD-ROM.

 Why do freshwater sponges have gemmules as part of their life cycle?

Cross-References to Other McGraw-Hill Zoology Titles

Integrated Principles of Zoology, 11th edition. C. P. Hickman, L. S. Roberts, and A. Larson. Chapter 12.

Animal Diversity, 3rd edition. C. P. Hickman, L. S. Roberts, and A. Larson. Chapter 6.

Zoology, 5th edition. S. A. Miller and J. P. Harley. Chapter 9.

Biology of the Invertebrates, 4th edition. J. Pechenik. Chapter 4.

Laboratory Studies in Integrated Principles of Zoology, 10th edition. C. P. Hickman, F. Hickman, and L. Kats. Chapter 7.

General Zoology Laboratory Guide, 13th edition. C. Lytle. Chapter 6.

Structures Checklist

Here are some of the structures that you should be able to easily find in *Digital Zoology* and the specimens that you will be looking at in your lab. After reading your lab handout, you might want to add more and, depending on the equipment available in your lab, you might see more. As you study the material, you might also want to make some notes on how some of these structures looked or include a drawing in your lab notes. (Structures indicated by * may be hard to see.)

Asconoid sponge: **Leucosolenia**

Preserved specimen
☐

☐ Body wall with spicules
☐ Osculum
☐ Spongocoel
☐

Cross section

☐ Body wall with spicules
☐ Choanocyte surface
☐ Porocytes*

☐ Spongocoel
☐
☐
☐
☐
☐

☐

Additional structures

☐
☐
☐
☐

Syconoid sponge: **Grantia** *or* **Scypha**

Preserved specimen

☐ Apopyles
☐ Body wall
☐ Incurrent canals
☐ Osculum
☐ Radial canals
☐ Spicules
☐ Spongocoel
☐

☐
☐

Cross section

☐ Amphiblastula larvae*
☐ Apopyles
☐ Choanocyte flagella*

☐ Choanocytes
☐ Incurrent canals
☐ Mesohyl
☐ Prosopyles
☐ Radial canals
☐ Spicules
☐ Spongocoel
☐
☐

☐

Additional structures

☐
☐
☐
☐
☐
☐

Sponges, spicules, and gemmules

☐ Gemmule
☐ Hexaxon
☐ Monoaxon

☐ Polyaxon
☐ Spongin
☐ Triaxon

☐ Triradiates
☐
☐

Crossword Puzzle – Porifera

Puzzle solution. An interactive web-based version of this puzzle and its solution are available on the *Digital Zoology* website www.digitalzoology.com/students. With the interactive puzzle you can check to see if individual words or the whole puzzle is correct and get hints for single letters.

Across

2 The flagellar cells of the developing larval scyphan sponge move toward the outside in a process called? (9)

4 Compared to asconoid and syconoid sponges, the leuconoid sponges have _____ oscula (osculum). (8)

8 Describes the plasticity of sponge cells and how different seemingly specialized types can change into others. (10)

10 The microvilli of these cells form this part of the choanocyte. (6)

11 This cavity is lined with choanocytes in asconoid sponges but they're missing in the same cavity in the syconoid form. (10)

13 The name for the unique six-pointed spicule of glass sponges. (7)

14 Canals in a syconoid sponge; the choanocytes are found here. (6)

15 This dormant sponge structure helps a freshwater sponge get through the winter. (7)

17 The number of different sponge architectures. (5)

19 The opening, in all types of sponge architectures, that water pumped by the sponge exits. (7)

22 In leuconoid sponges, the choanocytes are found inside these. (8)

24 Describes multicellular animals that lack tissues. (7)

25 Specialized types of these cells transport sperm caught by the collar cell to the egg in a sponge. (11)

28 As water moves through the choanocyte chambers it does this so that it can be filtered. (5)

30 These large cells found in the larval stage of some sponges will form the outer choanoderm. (10)

31 If you were a sponge taxonomist you would spend a lot of your time examining these. (8)

Down

1 Water enters an asconoid sponge through this special sponge cell. (8)

3 A proteinaceous sponge spicule. (7)

5 The name for the outer layer of cells in a sponge. (10)

6 Sponges are organized at this grade. (8)

7 Sponges are called this type of feeder because they do this to the water that they pump. (6)

9 The taxonomic level of the Porifera. (6)

11 The sponge cell that produces the spicules. (10)

12 Unlike a leuconoid sponges, asconoid and syconoid sponges have how many oscula? (3)

15 Reduction bodies in sponges are also called this. (8)

16 Sponge spicules can be grouped into two types, the small microscleres that reinforce or pack of the sponge body and these larger spicules. (12)

18 Sponge architecture consisting of choanocytes in the spongocoel and a single osculum. (8)

20 An amphiblastula is this stage in the life cycle of some sponges. (6)

21 The structure on a choanocyte that propels the water through a sponge. (9)

22 Although it's not a tissue, because sponges don't have them, the cells lining the spongocoel are collectively called this. (10)

23 This hollow larval stage of some sponges resembles the developmental stage in higher animals. (8)

26 Sponges are organized at the cellular grade and don't have cells organized as these. (7)

27 Compared to the hollow coeloblastula, the parenchymula larva in a sponge is best described as this. (5)

28 Sponges found in the deepest parts of the oceans have spicules made of this. (6)

29 Sponges are becoming of interest because not only do they protect themselves with spicules but also produce these to defend themselves. (6)

Self-Test - Porifera

Use the following labels to identify the photographs. You may have to use a label more than once, and some labels may not be appropriate for the photographs. Answers are available on the *Digital Zoology* website at www.digitalzoology.com/students. Be sure to try the interactive drag-and-drop quizzes on the *Digital Zoology* CD-ROM.

Specimens

A- *Grantia (Scypha)*

B- *Spongilla*

Labels

1) Apopyle
2) Choanocyte chamber
3) Choanocytes
4) Gemmule
5) Incurrent canal
6) Mesohyl
7) Osculum
8) Porocyte
9) Prosopyle
10) Radial canal
11) Spicule
12) Spongocoel

CNIDARIA

Inside *Digital Zoology*

As you explore the Cnidaria on the *Digital Zoology* CD-ROM or on the web at www.digitalzoology.com, don't miss these learning tools:

- ○ Photos of preserved specimens and cross sections of hydrozoans including: *Hydra; Obelia; Gonionemus; Pennaria;* and the Portuguese Man-of-War, *Physalia.* The jellyfish *Aurelia,* represents the Schyphozoa and the sea anemone *Metridium* the anthozoan body plan. Take a look at a variety of different corals and explore the various stages in the cnidarian life cycle with the interactive life cycles.
- ○ Video of *Hydra* and how it moves and changes shape.
- ○ Drag-and-drop quizzes on the key structures of *Hydra, Aurelia,* and *Metridium.*
- ○ Interactive cladogram showing the major events that gave rise to the four cnidarian classes. A summary of the key characteristics of each class are combined with an interactive glossary of terms.

Defining Differences

Cnidarians are an important group for us to study because their unique features are intermediate between the first multicellular organisms that preceded them, the Porifera, and those that followed. Some of the features in the following list appear for the first time in the Cnidaria, and they define the phylum. Other characters in the list are things that we see for the first time in animals and are important for understanding how organisms in the phylum function. Whichever is the case, you'll want to watch for examples of these in the various cnidarian specimens that you will be examining.

- ○ Radially symmetric animals organized at the
- ○ diploblastic tissue-grade organization with unique
- ○ cnidocytes, an
- ○ incomplete gut that can be used as a
- ○ hydrostatic skeleton, and a
- ○ dimorphic life cycle.

Radially symmetric

Radially symmetric animals do not consistently orient in any particular direction and don't have an anterior or posterior end like **bilaterally symmetric** animals do. Instead, cnidarians have oral and aboral surfaces. One of the most distinctive features of the Cnidaria is their radial symmetry, and it gives us some clues to how they function.

One consequence of the absence of directed movement is that these animals have a nerve net distributed throughout the body rather than being complex and centralized at the anterior end of the animal, which is the case in bilaterally symmetric animals. If you don't always face or move in the same direction, there's no advantage to concentrating sensory structures to detect what is ahead. Instead, sensory structures are evenly spaced around the whole organism allowing it to react equally well in all directions. Watch for the position of **rhopalia,** tentacles, and any other structures radially arranged in the specimens you look at in your lab and in *Digital Zoology.*

One group of cnidarians take radial symmetry a step farther. The anemones and corals (Anthozoa) show **biradial symmetry**. Even though most of their structures are radially arranged around the oral-aboral axis, there is at least one important paired structure that restricts the planes of symmetry to only two. When you look at the anemone watch for the **syphonoglyphs** and the arrangement of the internal **septa** to see why these animals are biradial, rather than radial.

One of the assumptions often made about **primitive** and **advanced** traits in the evolution of animals is that radial symmetry is primitive and bilateral is advanced. It's based on the idea that radially symmetric animals are **pelagic**, passively floating in their marine environments collecting food as they bump into it, or sessile, sitting on the bottom waiting for food to float by. Pelagic animals became sessile, then started to move across the substrate. The anthozoan's shift to biradial symmetry could be an intermediate step on the way to the more advanced bilateral symmetry.

Not everyone agrees, and an alternate would see the first animals being motile and bilaterally symmetric with some adopting a sessile existence and the benefits of radial symmetry for that type of life. It's not all that far fetched. We will see later that part of the sequence is exactly what the echinoderms did.

 What other types of symmetry can animals have?

Diploblastic tissue grade organization

The cnidarians are the flip side to the multicellular coin, being the first animals to have tissues derived from **endoderm** and **ectoderm**. They only have two of the possible three germ layers, and that's why we refer to them as **diploblastic**. The next change we'll see is the presence of the third germ layer, the **mesoderm**, and being **triploblastic** with all three cell layers.

These animals are organized at the **tissue grade** because neither ectoderm nor endoderm have been modified to form organ systems. That doesn't mean their tissues are simple, with all cells doing the same thing. There is division of labor between the cells. A good example of this can be seen in the various cells found in the outer **epidermis** and the inner gastrodermis.

 What's the difference between a mesoglea and a mesenchyme found between the two layers of a diploblast?

The outer surface of the animal is defined by the **epitheliomuscular cells** that connect with each other to form the body covering. As you might have guessed from their name, these cells do more than that. They also have contractile myonemes at the base of the cell, and these are aligned in the same direction. Nerve cells, also formed from the ectoderm, wind between the epitheliomuscular cells to create the nerve net.

Endoderm cells form the **gastrodermis** that lines the **gastrovascular cavity** (**coelenteron**), and they, too, have a variety of functions. For example, **nutritive muscular cells** are responsible for phagocytic uptake of food from the gastrovascular cavity. Their cell bases also have myonemes arranged in the opposite direction to those of the epitheliomuscular layers. Another type of cell found in the gastrodermis are gland cells, which secrete digestive enzymes into the gastrovascular cavity to start the digestion of ingested food.

Cnidocytes

The most specialized cell found in a cnidarian gives the phylum its name. These are the stinging **cnidocytes** embedded in the epidermis around the mouth and, when present, the tentacles, or arms. Cnidarians are predators and capture their prey using these unique stinging cells. The secret to how they work is the cell's specialized organelle, the **nematocyst**, found near the cell surface. Coiled inside the cnidocyte's nematocyst is a deadly barbed, sticky, or poison-tipped, thread waiting to be fired if the cell is triggered. When fired, they shoot out at tremendous velocities easily penetrating, or entangling, potential prey. Once captured, prey is put inside the gastrovascular cavity where it is digested.

Incomplete gut

The digestive tract of a cnidarian is an **incomplete gut** because it has only one of the two possible openings that guts have–a mouth, but no anus. Once food is in the gut, the mouth closes, and the cells of the gastrodermis begin to work. Digestive enzymes are secreted into the gastrovascular cavity. The first part of the digestive process breaks down the food into particles small enough for the nutritive muscular cells to consume by phagocytosis–the final stage of the digestive process.

That's the simple view of the digestive system. A lot of cnidarians are colonial, and each member of the colony is connected to the next by extensions of the gastrovascular cavity. The flagellated cells of the gut ensure that food is distributed to everyone in the colony using these common connections. Jellyfish have a different set up, and their gut isn't as much a cavity as a series of canals that allow the nutrients to flow and reach all the different parts of the animal.

There is at least one disadvantage to this type of gut–any undigested food must be emptied from the gut through the mouth.

Hydrostatic skeleton

Cnidarians use their gut for more than digestion, and we see this new functional system for the first time in the phylum, a skeleton–a **hydrostatic skeleton**. When a cnidarian's mouth is closed, the gastrovascular cavity is a closed fluid-filled space, and if muscles surrounding the cavity contract, they'll change the shape of the space.

Think of it as a water-filled cylindrical balloon with two layers of contractile cells, one with the myonemes running the length (longitudinal) of the balloon and the other arranged in circles (circular) around the balloon's diameter. When the longitudinal ones contract, they shorten and the diameter of the balloon increases. The opposite happens when the circular myonemes contract. The balloon becomes a longer and narrower cyclinder. But there is more than just a change in shape going on, and it's why we call this a skeleton.

Myonemes, or muscles, burn energy to contract and shorten, but they have no way to return to their original length unless something stretches them. That's the role of the skeleton with its two sets of antagonistic muscles. When one contracts it causes the other to lengthen, and in a hydrostatic skeleton the muscles use fluid to do that. That increasing diameter that comes from contraction of the longitudinal myonemes stretches the circular muscles back to their original length. The result of this is that these changes in shape allow for movement.

Dimorphic life cycle

As it changes from a sessile polyp to the mobile medusa, a cnidarian undergoes a tremendous transformation in its appearance and how it functions. It's why their life cycle is dimorphic, and for the first cnidarians, both stages were equally important to how the animal lived. As the group evolved, the importance of the two stages changed. In some cnidarians, the animal's entire life is spent as a polyp; in others, as a medusa; and in still others, something in between.

 What are some of the different roles the polyp and medusa have in the different cnidarian classes?

The larval stage of the cnidarians is a planula, a ciliated solid ball of cells that forms from the fertilized egg. The planula is motile, and after swimming through the water for a while it settles down on the substrate and develops into the polyp. The cnidarian planula is often used in theories of the origins of the metazoa, so be sure to take a look.

 Why is the planula important in theories of metazoan evolution?

A Closer Look Inside *Digital Zoology* - Cnidaria

Hydrozoa: Obelia *and* Hydra

First appearances can be deceiving. A quick look at the preserved specimen of *Obelia* and you might think you are looking at some sort of plant. (If you don't have a preserved specimen, take a look inside *Digital Zoology* for an example.) If you have a preserved specimen, take a closer look under the microscope and you'll see the different **zooids** (hydroids) that cover the stalks. Even with the preserved specimen you may find it hard to see any details of the zooids because they are covered with a nonliving protective **perisarc**. In prepared slides, the living tissue, **coenosarc**, is stained and you can see it through the perisarc of the animal that makes up the colony.

Almost all hydrozoans form some sort of colony, and polyps in the colony often have different jobs to do–another example of division of labor. The two most common jobs for polyps are feeding and reproduction. The **gastrozooids** have tentacles and feed, and the **gonozooids** produce the small medusa that will grow into the reproductive free-swimming medusa. Each of the polyps is connected to the other through the stems and branches of the colony.

Hydra is often used in the zoology labs to show the various features of a cnidarian polyp, and it's a great specimen for doing that. What may be confusing is that this hydrozoan's life cycle is spent entirely in the polyp stage; there's no medusa. So why is it a hydrozoan? It's important to remember that ancestral characteristics of a group can be modified in a variety of ways, and the original character often doesn't look at all like the trait when it first appeared. The most drastic modification, a trait's complete

---

disappearance, is an example of an advanced condition. An absence of the medusa doesn't disqualify *Hydra* from being a member of the Hydrozoa. With the reproductive medusa gone, the polyp has reproductive responsibilities, and you'll see examples of both asexual and sexual reproduction in *Digital Zoology* and your lab materials.

If they aren't available in your lab, *Digital Zoology* has a variety of other colonial hydrozoans at which you can take a look. One of these is *Gonionemus,* where you'll see the velum, also a hydrozoan trait.

✍ *Why is the velum important in how the hydrozoan medusa swims?*

The most spectacular colonial hydrozoan is the Portuguese man-of-war, *Physalia*. These large hydrozoans drift in the oceans using a modified polyp filled with air that acts as both a float and sail. Inflated with gas, it keeps the colony at the water surface, and as the winds blow against the part that's above the water line, the same modified polyp acts a sail. Dangling behind are a variety of tentacles armed with cnidocytes. Clustered among these, are the feeding and reproductive polyps of the colony.

✍ *How many different types of polyps does the Portuguese man-of-war have?*

Scyphozoa: *The jellyfish* Aurelia

The jellyfish life cycle, like all scyphozoans, has the usual cnidarian stages. The difference is that in this class the polyp is highly modified and only around for a short time when compared to the longevity of the medusa.

Aurelia's gonads are located next to the four gastric pouches, and gametes are released into the gastrovascular cavity and out the mouth. The planula swims around for a while before it settles down on the substrate, and changes into the unusual polyp stage, the **scyphostome**. This grows and then undergoes transverse fission creating the **strobila,** which look like a stack of dinner plates piled one on top of the other. Each of the "plates" develops small arms and then separates from the top of the strobila as an **ephyra,** which ultimately grows into the mature adult jellyfish whose shape with which you might be more familiar.

As jellyfish float through the water, small zooplankton are trapped in mucus on the underside of the bell where cilia, along with the help of the oral arms, move the food into the mouth and the gastrovascular cavity. Compared to a polyp, the gastrovascular cavity is extensively modified, and the mouth leads to four gastric pouches connected to a complex series of canals that lead to the radial canal around the margin of the bell. These canals work like a circulatory system, and food follows a well-defined path as it moves through the different parts of the bell of the medusa and distributes nutrients to the tissues.

What moves the fluid in a jellyfish's gastrovascular cavity?

Anthozoa: **The anemone** *Metridium*

If you've been working your way through the invertebrates phylogenetically in either your lab or *Digital Zoology*, you're probably glad to finally get to look at a decent-sized animal like the sea anemone. But remember, this is one of the two different types of anthozoa, and the small-size delicate corals–also anthozoans–have a huge impact on the marine environment building of coral reefs, some of the most productive marine environments. The most striking thing about anemones is their size, especially for a diploblastic organism. One of things you'll want watch for as you work with the specimen are adaptations that allow these animals to be this big.

The mouth is surrounded by tentacles that trap food, and then push it into the gastrovascular cavity where it is digested. Take a close look at the mouth and you'll see that it isn't a perfect circle; it's more oval in shape. At the edges of the mouth are paired ciliated grooves, the **siphonoglyphs,** which move fresh oxygenated water into the gastrovascular cavity, and that's important for the complex of tissues found inside the cavity. The presence of these functionally important paired structures creates a new type of symmetry that we haven't seen before, biradial symmetry.

Internally, the gastrovascular cavity is divided by either **incomplete septa** or **complete septa** that increase the surface area available for digestion of food. It's the incomplete septa that are important for digestion. The free ends of the septa have three lobes where there are glandular cells that release the digestive enzymes, ciliated cells to mix the content of the cavity, and cnidocytes. Anemones consume large prey, and having cnidocytes inside the digestive tract is a big help in killing prey that may still be alive when ingested. Anthozoans are the only cnidarians with cnidocytes in the gastrovascular cavity.

Complete septa do a little more than connect with the pharynx. They also have retractor muscles that help change the shape of the anemone, mix the contents of the gastrovascular cavity, and help to move the tentacles that surround the mouth. If you have trouble telling the difference between these different parts of the cavity in your dissected specimen, take a look at the cross section slide if it's available in your lab, or take a look in *Digital Zoology*.

 If cnidarians are diploblastic, lacking mesoderm, how can an anemone have retractor muscles?

Cross-References to Other McGraw-Hill Zoology Titles

Integrated Principles of Zoology, 11th edition. C. P. Hickman, L. S. Roberts, and A. Larson. Chapter 13.

Animal Diversity, 3rd edition. C. P. Hickman, L. S. Roberts, and A. Larson. Chapter 7.

Zoology, 5th edition. S. A. Miller and J. P. Harley. Chapter 9.

Biology of the Invertebrates, 4th edition. J. Pechenik. Chapter 6.

Laboratory Studies in Integrated Principles of Zoology, 10th edition. C. P. Hickman, F. Hickman, and L. Kats. Chapter 8.

General Zoology Laboratory Guide, 13th edition. C. Lytle. Chapter 7.

Structures Checklist

Here are some of the structures that you should be able to easily find in *Digital Zoology* and the specimens that you will be looking at in your lab. After reading your lab handout, you might want to add more and, depending on the equipment available in your lab, you might see more. As you study the material, you might also want to make some notes on how some of these structures looked or include a drawing in your lab notes. (Structures indicated by * may be hard to see.)

Obelia

Hydroid colony

- ☐ Blastostyle
- ☐ Coanosarc
- ☐ Gastrovascular cavity
- ☐ Gastrozooid (hydranth)
- ☐ Gonotheca

- ☐ Gonozooid (gonangia)
- ☐ Hydrocalus
- ☐ Hydrorhiza
- ☐ Hydrotheca
- ☐ Hypostome
- ☐ Medusa buds
- ☐ Mouth
- ☐ Perisarc
- ☐ Tentacles

- ☐
- ☐
- ☐

Medusa

- ☐ Manubrium
- ☐ Medusa
- ☐ Nematocysts
- ☐ Radial canals
- ☐ Statocysts

- ☐ Tentacles
- ☐
- ☐

Additional structures

- ☐ Planula
- ☐

Hydra

- ☐ Basal disc
- ☐ Buds
- ☐ Cnidocytes on tentacles
- ☐ Epidermis
- ☐ Epitheliomuscular cells*

- ☐ Gastrodermis
- ☐ Gastrovascular cavity (Coelenteron)
- ☐ Gland cells*
- ☐ Hypostome
- ☐ Interstitial cells*

- ☐ Mesoglea
- ☐ Mouth
- ☐ Nutritive-muscular cells*
- ☐ Ovaries
- ☐ Spermaries (testes)
- ☐ Tentacles

- ☐
- ☐

Additional structures

- ☐ Planula
- ☐

Aurelia

Adult

- ☐ Gastric pouches
- ☐ Gonads
- ☐ Lappet
- ☐ Marginal tentacles
- ☐ Oral arms
- ☐ Oral tentacles
- ☐ Radial canal
- ☐ Rhopalia
- ☐ Ring (circular) canals
- ☐ Stomach
- ☐
- ☐
- ☐
- ☐

Life cycle

- ☐ Ephyra
- ☐ Planula
- ☐ Schyphistome
- ☐ Strobila
- ☐
- ☐
- ☐

Additional structures

- ☐
- ☐
- ☐

Gonionemus

- ☐ Cnidocytes
- ☐ Gonad
- ☐ Manubrium
- ☐ Mouth
- ☐ Radial (circular) canal
- ☐ Ring canal
- ☐ Tentacles
- ☐ Velum
- ☐
- ☐

Additional structures

- ☐

Metridium

Cross section

- ☐ Ciliated lateral lobe of septa
- ☐ Cnidoglandular lobe of septa
- ☐ Complete septa
- ☐ Epidermis
- ☐ Gastrodermis
- ☐ Gastrovascular cavity (coelenteron)
- ☐ Incomplete septa
- ☐ Mesenchyme (mesoglea)
- ☐ Pharynx
- ☐ Retractor muscles
- ☐ Siphonoglyphs
- ☐
- ☐
- ☐

Preserved specimen: Intact and dissected

- ☐ Acontia filaments
- ☐ Complete septa
- ☐ Gastrovascular cavity (coelenteron)
- ☐ Incomplete septa
- ☐ Mouth
- ☐ Oral disc
- ☐ Pedal disc
- ☐ Pharynx
- ☐ Retractor muscle
- ☐ Siphonogyphs
- ☐ Tentacles
- ☐
- ☐
- ☐

Additional structures

- ☐
- ☐
- ☐
- ☐
- ☐

Crossword Puzzle - Cnidaria

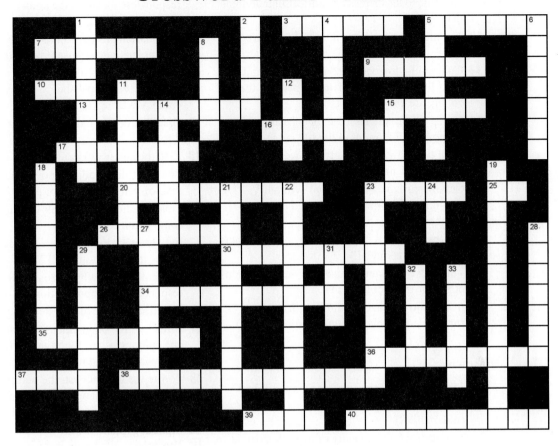

Puzzle solution. An interactive web-based version of this puzzle and its solution are available on the *Digital Zoology* website www.digitalzoology.com/students. With the interactive puzzle you can check to see if individual words or the whole puzzle is correct and get hints for single letters.

Across

3 The type of symmetry characteristic of cnidarians. (6)

5 Although it is a hydrozoan, *Hydra* is missing this stage of the typical hydrozoan life cycle. It's thought to be an adaptation to living in fresh water. (6)

7 A true nematocyst injects proteins and this toxic compound when it penetrates its prey. (6)

9 Cnidarians bodies are organized at this grade; they have these but still don't have organ systems. (6)

10 Biradially symmetric anthozoans have this many syphonoglyphs; it's one of the reasons they have biradial rather than radial symmetry. (3)

13 This nonliving structure surrounds the gonozooid of a thecate hydrozoan. (9)

15 In a scyphozoan, the schyphostome is all that remains of this stage usually found in the cnidarian life cycle. (5)

16 Epitheliomuscular cells are found on this side of the cnidarian body wall. (7)

17 Gastrozooids have this responsibility in a hydrozoan colony. (7)

20 This cnidarian cell type ultimately develops into a cnidocyte. (10)

23 These cells secrete digestive enzymes into the digestive cavity of a cnidarian. (5)

25 Is the perisarc in a colonial cnidarian living material? (2)

26 This ciliated solid ball, a cnidarian larva, may be a clue to the origins of multicellular animals. (7)

30 The medusa dominates in this cnidarian's life cycle. (9)

34 Colonial hydrozoans resemble plants, and the roots that connect the different parts of the colony are called this. (10)

35 In this cnidarian class, the medusa and polyp are equally important in the life cycle. (8)

36 With two different body types, we refer to the life cycle of cnidarians as being this. (9)

37 Cubozoans have quadraradial symmetry, meaning their symmetry is based on this number. (4)

38 Defensive zooids are technically referred to as this type of zooid. (13)

39 This part of a plant is most like the hydrocalus of a hydrozoan colony. (4)

40 In thecate hydrozoans the gastrozooid is surrounded by this nonliving structure. (10)

Down

1 The layer, remember it's not a tissue, found between the inner and outer tissue layers of a cnidarian. (8)

2 The gastrovascular cavity of an anthozoan is divided by these. (5)

4 Dactylozooids are specialized for this. (7)

5 One of the consequences of not having any mesoderm is that you don't have these either. (7)

6 Located at the base of an anemone's (Anthozoa) gastrovascular cavity, these help kill the ingested prey. (7)

8 The lobe in an anthozoan's incomplete septa where you would find the cnidocytes. (5)

11 Like the trigger of a gun, this structure is responsible for firing the cnidarian's cnidocyte. (8)

12 A cnidarian gastrovascular cavity is an incomplete gut because it lacks this. (4)

14 The energy to fire the nematocyst forms as the organelle develops--the _____ hypothesis for cnidocyte firing. (7)

15 Disk that corals and other anthozoans sit on this. (5)

18 A statocyst always contains one of these. (9)

19 The gas-filled modified polyp in the Portugese man-of-war is this type of polyp. (13)

21 Small medusa form on this central rod found in some gonozooids. (11)

22 In the scyphozoan life cycle, once the planula settles it turns into this. (12)

23 These polyps in a hydrozoan colony are responsible for reproduction. (10)

24 Describes the nervous system in a cnidarian. (3)

27 Unlike all the other cnidarian classes, in this one cnidocytes are found inside the gastrovascular cavity. (8)

28 To fire the nematocyst, the permiability of the cnidocyte cell changes in some way. This is a process referred to as this hypothesis for cnidocyte firing. (7)

29 This embryonic tissue layer is missing in a diploblast. (8)

31 If you are looking straight into the mouth of a hydrozoan, this surface is facing you. (4)

32 As the medusa contracts, this structure decreases the opening through which water is expelled. It's also a characteristic of a hydrozoan medusas. (5)

33 This free-swimming stage in the schyphozoan life cycle matures into an adult medusa. (6)

Self-Test - Cnidaria

Use the following labels to identify the photographs. You may have to use a label more than once, and some labels may not be appropriate for the photographs. Answers are available on the *Digital Zoology* website at www.digitalzoology.com/students. Be sure to try the interactive drag-and-drop quizzes on the *Digital Zoology* CD-ROM.

Specimens

A- *Hydra*
B- *Aurelia*
C- *Obelia*
D- *Gonionemus*
E- *Metridium*

Labels

1) Bud
2) Coenosacrc
3) Epidermis
4) Gastric pouch
5) Gastrodermis
6) Gastrovascular cavity
7) Gastrozooid
8) Gonad
9) Gonotheca
10) Gonozooid
11) Hypostome
12) Incomplete septa
13) Mesoglea
14) Oral arm
15) Ovary
16) Perisarc
17) Retractor muscle
18) Rhopalium
19) Tentacle
20) Velum

PLATYHELMINTHES

Inside *Digital Zoology*

As you explore the Platyhelminthes on the *Digital Zoology* CD-ROM or on the web at www.digitalzoology.com, don't miss these learning tools:

❍ Photos of whole mount and cross section slides of free-living turbellarians *Dugesia* (planaria) and whole mounts of *Bdelloura*. Trematode flukes include the Chinese liver fluke *Opisthorchis (Chlonorchis) sinensis*, sheep liver fluke *Fasciola hepatica*, *Paragonimus macrorchis* and *Schistosoma mansoni*. Cestodes are represented by *Diplydium caninum* and *Taenia pisiformis*. Different stages of some of the parasitic flatworms can be explored using interactive life cycles.

❍ Video clips of flatworm movement showing gliding ciliary movement and use of musculature in *Dugesia* are available in addition to video clips showing the avoidance response and the pharynx.

❍ Drag-and-drop quizzes on the key structures of the *Dugesia*, the Chinese liver fluke, and tapeworms.

❍ Interactive cladogram showing the major events that gave rise to the four platyhelminth classes. A summary of the key characteristics of each class are combined with an interactive glossary of terms.

Defining Differences

Some of the differences described in the following sections appear for the first time in the Platyhelminthes, and they define the group. Others are important for understanding how animals in the phylum function. Whichever is the case, you'll want to watch for examples of these in the various flatworm specimens that you will be examining.

❍ Bilaterally symmetry,
❍ triploblast acoelomates, at the
❍ organ grade organization that have either
❍ free-living or parasitic lifestyles.

Bilateral symmetry

Bilateral symmetry appears for the first time in the flatworms. The common assumption is that this type of symmetry is more advanced than the preceding **radial symmetry**, or **biradial symmetry.** (This is discussed in more detail in the previous chapter on Cnidaria.) Bilaterally symmetric animals have anterior and posterior ends, left and right sides, as well as dorsal and ventral surfaces. But being bilateral is more than just a new term to describe the parts of the animal. There are advantages to being able to sense where you're going, and there is increasing complexity of the nervous system and a concentration of sensory structures at the anterior end of the animal. It's a process referred to as **cephalization**, a first with the flatworms, but with a difference. In other cephalized animals, the mouth is located at the anterior end but that's not the case in flatworms, and their mouth is located on the ventral surface.

Describe the cephalization of the flatworm's nervous and sensory systems.

Triploblast acoelomates

The Platyhelminthes are the first **triploblastic** animals and have a third tissue layer, the **mesoderm**, added between the **ectoderm** and **endoderm** that cnidarians had. Flatworms, and all the other triploblasts, can now have muscles, formed from mesoderm. Platyhelminthes don't realize the full potential of the triploblast body plan and, unlike the other animals to follow, a flatworm's mesoderm is solid without a cavity, a **coelom**, inside. It's why they are referred as **acoelomate**. About the only way a solid animal can maintain an adequate surface-to-volume ratio is to be as flat as possible, and that's what these animals do. It explains how the phylum got its common name, flatworms.

Solid organisms have to depend on diffusion to move oxygen, nutrients, and metabolic wastes, and the same is true for flatworms. Like the cnidarians, the digestive system is incomplete, but branches throughout the body making sure that every cell is close to a supply of nutrients. The gut contents are mixed by the cilia that line it and also by the movements of the animal. The fluids in the gut and the surrounding environment are important for gas exchange and removal of metabolic wastes. Flatworms also have **protonephridia** for **osmoregulation** and are most obvious in freshwater species, which are **hyperosmotic** to the water around them. Protonephridia are also called flame cells, and they collect the interstitial fluid found between the cells and remove it from the animal.

How does a flame cell work?

Organ-grade organization

Even with the limitations of being acoelomate, flatworms have developed complex organ systems including: nervous and excretory systems and a complicated reproductive system, easy to see in your lab specimens and in *Digital Zoology*. Flatworms are **monoecious**, and each animal has both male and female reproductive tracts–they're hermaphrodites, and that's why the reproductive system appears to be so complicated.

Compared to **asexual reproduction**, one of the advantages of **sexual reproduction** is that there is mixing of genetic material from two different parents, and this results in offspring that have a different genetic make up than their parents. Being monoecious doesn't mean that this advantage is lost. Flatworms, like all monoecious organisms, go to great lengths to be sure that they don't fertilize their own eggs.

The secret is that events of insemination(transferring the sperm) and fertilization (combining the sperm and egg) have been separated from each other. Here's how it works. In each flatworm, sperm is formed in the **testes**, then transported by the **vas deferens** (sperm ducts) to the **seminal vesicles** where it is stored until the animal meets with another flatworm. When the two mate, each passes sperm from their own seminal vesicle to the other flatworm, who stores the sperm that they received in their **seminal receptacle**. Now that insemination is over, the two mating flatworms separate and head off on their own.

Some time later fertilization occurs. Eggs produced in the **ovaries** pass down the **oviduct** and are filled with nutritious yolk from the **vitelline glands**. The eggs then pass by the opening of the seminal receptacle and are fertilized using the stored sperm from the other flatworm. The result is that the flatworms that hatch from these fertilized eggs have genetic material from each of the parents, and variation is possible.

✍ *What advantage could there be to being a monecious organism?*

Free-living or parasitic lifestyles

In the evolutionary sequence, flatworms are replaced by coelomate animals, and it's not unlikely that the new coelomates fed on flatworms. A flatworm gliding by on the substrate might be a tasty morsel for a potential coelomate predator, and living flatworms give us clues to how the early animals in the phylum survived this new threat to still be around today. They either defended themselves against predation or moved into the nice fluid-filled cavity of these new coelomate animals, living there instead.

Ancestral flatworms moved using cilia that beat in a layer of mucus that the epithelium lay down on the substrate. Any modifications of the mucus that were distasteful to would-be predators were advantageous, and flatworms still use this defensive strategy. In some flatworms **rhabdites** form in the epithelium, and when these rod-shape structures are discharged, they rapidly dissolve to make the distasteful mucus that protects the flatworm. One of the more unusual defenses is found in flatworms that feed in cnidarians. Cnidocytes that the flatworm consumes don't fire, and they're moved to the flatworm's epithelium and used for defense.

Another way that flatworms move is to stick to the substrate, and then use contractions of the longitudinal and circular muscles to move the body. Again, epithelial secretions are important in sticking to the substrate, and in parasitic flatworms dual adhesive glands are located in specific areas creating adhesive suckers used to stay attached to the host. In still other parasites, the epithelium secretes hooks used to hold the animal in place.

Adopting a parasitic existence is no easy task. To survive, a parasite can't kill its host, but has to be able to locate, and move into, a new host if the need arises. To do that, the number of parasites in the host needs to be increased. After all, if it's only a one in a thousand chance that the parasite can get to a new host, it may be out of luck if the one parasite that first found the host has only 500 offspring when it's time to move to a new host. Now, if there are 50,000 offspring, that's a different story. It's what the life cycles are about, increasing the numbers and getting to the next host. What makes the life cycles of parasitic flatworms stand out, especially those of the trematode flukes, is their complexity involving not one, but two and sometimes three, different animals.

What are the different stages in the fluke life cycle, and what does each contribute to the cycle?

A Closer Look Inside *Digital Zoology* - Platyhelminthes

Turbellaria: Free-living flatworms Dugesia *and* Bdelloura

The flatworm *Dugesia*, which has the common name planaria, is the most often studied example of a flatworm in zoology courses. If living specimens aren't available in your lab, be sure and see the two different ways that *Dugesia* moves in the video clips in *Digital Zoology*.

The first way of moving is by using cilia, and it results in a smooth gliding motion across the substrate. In the second, which can best be described as a "stick and stretch" technique, adhesive glands on its ventral surface are used to stick to the substrate. With the anterior end stuck in place, longitudinal muscles contract and *Dugesia* shortens its body. It then sticks the posterior end in place. Once that's done, the anterior end is released from the substrate. The circular muscles contract causing the body to lengthen and push against the anchored posterior end moving the anterior end forward. Once again stick the anterior back down and repeat the sequence.

Whole mount slides give you a good look at the digestive tract in these animals, and you'll see how it branches throughout the body to supply nutrients to all the cells and tissues. *Dugesia* is a **triclad** flatworm because its gut has three branches. You'll find the pharynx on the ventral surface, although you may have to flip your microscope slide upside down to see it. A short video clip in *Digital Zoology* shows the pharynx. Other easy features to see on whole mounts are the **eye spots** and auricles, which are evidence of cephalization and form part of the nervous system.

Prepared slides of cross sections usually include sections through the animal anterior and posterior to the pharynx and through the pharyngeal region. Be sure to look at each of these, and try and match where the section comes from with the whole mount slide that you looked at. As you look at your cross section slides be careful that you don't confuse the lumen of the digestive tract or the pharynx with a body cavity. Remember that these animals are acoelomate.

 How can you tell if your cross section is anterior to, posterior to, or through the pharyngeal region?

Differences in the length of the cilia make it possible to tell the dorsal surface, with its shorter cilia, from the ventral surface, with its longer cilia. Look toward the edges of the section, and you should be able to

---

see the adhesive glands that *Dugesia* uses to stick to the substrate. The circular and longitudinal muscles are easy to see along the edge of the body with a third type of muscle, the dorsoventral muscles, which extend from the dorsal to the ventral surface of the animal. These help to keep the flatworm flat as it contracts the other musculature.

If your lab doesn't have *Bdelloura* available, you might want to take a look at the whole mount of it in *Digital Zoology*. Preparations of this flatworm are more transparent than *Dugesia,* making it a lot easier to see the detail of the internal organs, especially the reproductive and nervous systems. *Bdelloura* is an ectoparasite found on the gills of horseshoe crabs. They stick to their host using a caudal sucker made of large adhesive glands similar to the ones found along the edges of *Dugesia.* When epidermal gland cells form a localized sticking structure like this, it's called a sucker, and we'll see more examples of this in the endoparasitic flatworms.

Trematoda: Trematode flukes

There are four different flukes available for you in *Digital Zoology*: the Chinese liver fluke *Opisthorchis (Chlonorchis)*, sheep liver fluke *Fasciola, Paragonimus,* and *Schistosoma.* The full life cycle of the Chinese liver fluke is also available. Photos of whole mounts of all four adults will allow you to identify various internal structures.

The Chinese liver fluke, *Opisthorchis*, is used in many labs to demonstrate the trematode life cycle. Slides of the adults allow you to easily see the mouth and digestive tract, which forks into two branches just behind the mouth. Most of the animal is filled with reproductive tract. Paired testes occupy the posterior end of the animal and the female reproductive tract, the middle and anterior. The ovary is in the middle along with the seminal receptacle where sperm from another fluke is stored. The uterus winds its way to the genital pore, and you should be able to see shelled eggs inside. Yolk glands on the sides of the animal connect in the region of Mehlis' gland and the opening of the ovary.

In the sheep liver fluke, *Fasciola*, the digestive tract makes it difficult to see many of the other structures. But if you look carefully you should be able to see most of the reproductive and digestive systems. Over all the organization is similar to that of the Chinese liver fluke with the male reproductive system occupying the posterior of the animal and the female in the anterior with the ovary near the middle. Vitelline glands are found down the sides. Try taking a close look at the anterior end of the animal so that you can differentiate what the digestive tract looks like from the reproductive system, which will help when you try and find the ovary. The uterus should be filled with shelled eggs, and posterior to it is the ovary. The vitelline duct carries yolk to the region of the ootype, an enlarged space in the reproductive tract, which also has openings from the ovary and seminal receptacle.

A third fluke that you'll find in *Digital Zoology* is *Paragonimus*. It's included in *Digital Zoology* because you'll have no trouble identifying the various parts of the internal anatomy with this species. Yolk glands, testes, ovaries, and the various tubes and ducts that connect them are easy to see in these almost transparent animals.

The last fluke we have for you is *Schistosoma*, which causes schistosomiasis. The adults of this species don't look like the other two. One of the big differences is that the sexes are separate in this species. Males and females also differ in their appearance with the males being the larger of the two and having a special groove that runs the length of his body, in which he holds a female.

In all four species be sure you can find both the anterior sucker and ventral sucker, characteristic of the trematodes.

Cestoda: Tapeworms

We have two different tapeworms for you to look at in *Digital Zoology*, and they closely resemble each other. In the dog tapeworm, *Dipylidium*, reproductive structures are paired with a gonopore on each side

---

of the **proglottid**. In the dog and cat tapeworm, *Taenia*, the structures aren't paired on each side. There are also morphological differences in the **scolex** that they use to attach to the intestinal wall of their vertebrate host.

The scolex is made up of hooks and suckers, and behind is a short neck that connects to the long chain of proglottids. New proglottids are added at the base of the neck, and the oldest ones are those farthest from the neck. Each proglottid is really just an egg-making machine, and when you examine the proglottids on the microscope slides, the reproductive tract is the most obvious feature. In younger proglottids you can see the testes, paired ovaries, and unpaired yolk gland. The vas deferens and oviduct both open in the gonopore. During mating sperm is passed out of the gonopore from the vas deferens and into the gonopore, oviduct, and to the seminal receptacle of the other tapeworm. Be sure to look at a mature, **gravid,** proglottid to see how it is nothing more than a sac of eggs. Eggs don't pass down the oviduct but are released when the proglottid bursts.

Tapeworms live in the digestive tract of their vertebrate host and are able to absorb nutrients from the digestive fluids that surround them. Tapeworms don't have a digestive tract of their own.

 How do tapeworms and flukes protect themselves from being digested or attacked by their host?

Cross-References to Other McGraw-Hill Zoology Titles

Integrated Principles of Zoology, 11th edition. C. P. Hickman, L. S. Roberts, and A. Larson. Chapter 14.

Animal Diversity, 3rd edition. C. P. Hickman, L. S. Roberts, and A. Larson. Chapter 8.

Zoology, 5th edition. S. A. Miller and J. P. Harley. Chapter 10.

Biology of the Invertebrates, 4th edition. J. Pechenik. Chapter 8.

Laboratory Studies in Integrated Principles of Zoology, 10th edition. C. P. Hickman, F. Hickman, and L. Kats. Chapter 9.

General Zoology Laboratory Guide, 13th edition. C. Lytle. Chapter 9.

Structures Checklist

Here are some of the structures that you should be able to easily find in *Digital Zoology* and the specimens that you will be looking at in your lab. After reading your lab handout, you might want to add more and, depending on the equipment available in your lab, you might see more. As you study the material, you might also want to make some notes on how some of these structures looked or include a drawing in your lab notes. (Structures indicated by * may be hard to see.)

Dugesia

Preserved or whole mount

☐
☐
☐

☐ Anterior branch of gut
☐ Auricles
☐ Eye spots
☐ Mouth
☐ Pharyngeal cavity
☐ Pharynx
☐ Posterior branches of gut

Cross section

☐ Anterior branch of gut
☐ Cilia
☐ Circular muscle
☐ Diverticula of gut
☐ Dorsoventral muscle

☐ Epidermis
☐ Gastrodermis
☐ Longitudinal muscle
☐ Lumen of pharynx
☐ Pharyngeal cavity
☐ Pharynx
☐ Posterior branch of gut
☐ Rhabdites*

☐ Ventral nerve cords*
☐
☐

Additional structures

☐
☐
☐

Bdelloura

Whole mount

☐ Adhesive disc
☐ Brain
☐ Diverticula of gut
☐ Eye spots
☐ Mouth

☐ Nerve cords
☐ Pharynx
☐ Seminal receptacles
☐ Sperm duct
☐ Testes
☐

☐
☐

Additional structures

☐
☐

☐
☐
☐

Opisthorchis *(Chlonorchis)*

Adult whole mount

☐ Anterior testis
☐ Bladder
☐ Eggs
☐ Esophagus
☐ Excretory duct*
☐ Excretory pore
☐ Genital pore
☐ Intestine

☐ Mehlis' gland
☐ Mouth
☐ Ootype
☐ Oral sucker
☐ Ovary
☐ Pharynx
☐ Posterior testis
☐ Seminal receptacle

☐ Sperm ducts (vas efferens & vas deferens)*
☐ Uterus
☐ Ventral sucker (acetabulum)
☐ Vitelline (yolk) duct*
☐ Vitelline (yolk) gland

☐
☐

Life cycle

☐ Cercaria
☐ Eggs
☐ Miracidium
☐ Redia inside sporocyst

Fasciola

Adult whole mount

- ☐ Anterior testis
- ☐ Eggs
- ☐ Esophagus
- ☐ Genital pore
- ☐ Intestine
- ☐ Mehlis' gland
- ☐ Mouth
- ☐ Ootype
- ☐ Oral sucker
- ☐ Ovary
- ☐ Penis*
- ☐ Pharynx
- ☐ Posterior testis
- ☐ Seminal receptacle
- ☐ Seminal vesicle*
- ☐ Sperm ducts (vas efferens & vas deferens)*
- ☐ Uterus
- ☐ Ventral sucker (acetabulum)
- ☐ Vitelline (yolk) duct*
- ☐ Vitelline (yolk) gland
- ☐
- ☐
- ☐
- ☐

Life cycle

- ☐ Cercaria
- ☐ Cyst
- ☐ Eggs
- ☐ Miracidium
- ☐ Redia inside sporocyst

Paragonimus

Adult whole mount

- ☐ Testes
- ☐ Eggs
- ☐ Esophagus
- ☐ Genital pore
- ☐ Intestine
- ☐ Mouth
- ☐ Ootype
- ☐ Oral sucker
- ☐ Ovary
- ☐ Pharynx
- ☐ Seminal receptacle
- ☐ Sperm ducts
- ☐ Uterus
- ☐ Ventral sucker (acetabulum)
- ☐ Vitelline (yolk) duct
- ☐ Vitelline (yolk) gland
- ☐
- ☐
- ☐
- ☐
- ☐

Dipylidium *or* Taenia

Adult whole mount

- ☐ Cirrus*
- ☐ Excretory canal
- ☐ Genital pore
- ☐ Ootype
- ☐ Ovary
- ☐ Oviduct
- ☐ Seminal receptacle*
- ☐ Sperm duct (vas deferens)
- ☐ Testes
- ☐ Uterus
- ☐ Vagina
- ☐ Yolk gland
- ☐
- ☐
- ☐
- ☐
- ☐

Additional structures

- ☐ Cysticercal stage
- ☐ Gravid proglottid
- ☐ Immature proglottid
- ☐ Neck
- ☐ Rostrum with hooks
- ☐ Scolex
- ☐ Suckers on scolex

Crossword Puzzle – Platyhelminthes

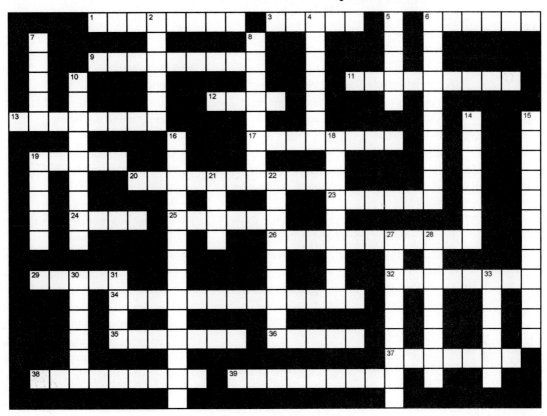

Puzzle solution. An interactive web-based version of this puzzle and its solution are available on the *Digital Zoology* website www.digitalzoology.com/students. With the interactive puzzle you can check to see if individual words or the whole puzzle is correct and get hints for single letters.

Across

1 This is the orientation of the outermost layer of muscles in free-living flatworms such as Planaria. (8)

3 Does an aquatic flatworm taste or smell it's environment? (5)

6 For cilia to work effectively the epidermis must first secrete this substance to cover the substrate that the flatworm glides across. (6)

9 This organ system is missing in tapeworms. (9)

11 This stage follows the miracidium in the fluke life cycle. (9)

12 Number of classes in the phylum Platyhelminthes. (4)

13 During the evolution of the different animal phyla, this tissue is found in the flatworms for the first time and is the reason they have muscles. (8)

17 Another name for the ocelli of free-living flatworms such as planaria. (8)

19 Because they resemble a flickering candle, this name applies to the cells of the protonephridia. (5)

20 These units repeat down the length of a tapeworm. (11)

23 The vas deferens transports gametes from this organ to the seminal vesicle. (6)

24 The name of the anterior-most sucker in a trematode. (4)

25 The junction of the oviduct, the opening of the seminal receptacle, and where the egg is usually fertilized in a flatworm. (6)

26 These very fine microvilli on a tapeworm's surface increase the surface area for absorption. (11)

29 The body of the ancestral flatworm is covered with these. (5)

32 These structures found on the sides of a free-living flatworm's head are chemosensory. (8)

34 The external opening of the protonephridia on the outer body wall of a flatworm. (13)

35 Females store their partners sperm in this receptacle. (7)

36 A flatworm in which the digestive system is a solid mass of endodermal cells, rather than a cavity. (5)

37 A flatworm with three main branches to its digestive tract is referred to as being this. (7)

38 Most species of flatworms have this type of life cycle. (9)

39 Another name for the yolk gland is this type of gland. (9)

Down

2 This cavity is missing in an acoelomate animal. (6)

4 When the adhesive glands in flatworms are located in one place helping to anchor it in place, they are referred to as these. (7)

5 Position of the epithelial cell bodies relative to the basement membrane in flatworm tegument. (5)

6 Flatworms are hermaphrodites; this is the more scientific term used to indicate this. (10)

7 The number of major muscle types, or layers, found in a free-living flatworm such as planaria. (5)

8 The special name for the protective body covering of parasitic flatworms. (8)

10 One of two types of hooks that you'll find on a monogenean fluke. (9)

14 This vital organelle lies below the plasma membrane of the tapeworm tegument and is protected. (7)

15 Unlike cestodes and trematodes, the monogenean flukes are this type of parasite. (12)

16 These are also called flame cells. (14)

18 The end of a monogenean fluke where you will find the opisthaptor. (9)

19 The common name for a trematode. (5)

21 How many anterior branches are there in a polyclad flatworm's gut? (4)

22 This flatworm class have acetabula; that's more than one acetabulum. (9)

27 Embedded in the epidermis, these protect a free-living flatworm from predation. (9)

28 Stage of a fluke life cycle that swims to either its vertebrate host or another host were it encysts. (8)

30 The nervous system of the flatworms looks like one of these. (6)

31 A flatworm's incomplete gut is missing this structure. (4)

33 The production of many cercaria from a single redia is an example of this type of amplification. (6)

Self-Test - Platyhelminthes

Use the following labels to identify the photographs. You may have to use a label more than once, and some labels may not be appropriate for the photographs. Answers are available on the *Digital Zoology* website at www.digitalzoology.com/students. Be sure to try the interactive drag-and-drop quizzes on the *Digital Zoology* CD-ROM.

Specimens

A- *Opisthorchis*
    (*Chlonochis*)
B- *Diplydium*
C- *Obelia*
D- *Paragonimus*
E- *Dugesia*

Labels

1) Adhesive gland
2) Anterior branch of triclad intestine
3) Anterior testis
4) Auricle
5) Cirrus
6) Diverticula of the gut
7) Epidermis
8) Excretory canal
9) Eye spot
10) Genital duct
11) Gonopore
12) Intestine
13) Ootype
14) Oral sucker
15) Ovary
16) Oviduct
17) Parenchyma
18) Pharyngeal cavity
19) Pharynx
20) Posterior branch of triclad intestine
21) Posterior testis
22) Seminal receptacle
23) Testis
24) Uterus
25) Vas deferens
26) Yolk (vitelline) gland

ANNELIDA

Inside *Digital Zoology*

As you explore the Annelida on the *Digital Zoology* CD-ROM or on the web at www.digitalzoology.com, don't miss these learning tools:

○ Photos of dissected specimens of both the preserved and anesthetized earthworms and preserved specimen *Nereis*.
○ Video clips of earthworm movement, internal organ movements including hearts and dorsal blood vessel, and pharynx.
○ Drag-and-drop quizzes on the key structures of the earthworm and *Nereis*.
○ Interactive cladogram showing the major events that gave rise to the three annelid classes.

Defining Differences

Some of the differences described in the following sections appear for the first time in the Annelida, and they define the phylum. Others are important for understanding how animals in the phylum function. Whichever is the case, you'll want to watch for examples of these in the various annelid specimens that you will be examining.

○ Metamerisation, a
○ a closed circulatory system; and
○ burrowing - a new form of locomotion.

Metamerisation

Unlike molluscs, where the coelom has been reduced and replaced with a hemocoel, the annelids retain a large and spacious body cavity. But annelids have done something a little different with the coelom. To become larger animals by repeating the coelomic unit, end to end, adding the new segments between the mouth and the anus. The proper term for the process is **metamerisation**, each of the segments is a **metamere**, and this is what give the annelids their unique segmented look.

In the ancestral annelid each metamere was a carbon copy of the one adjacent to it. Any structures formed from the **mesoderm**, and overlying **ectoderm**, were duplicated in each segment making each **serially homologous** to those adjacent to it. You could almost think of each segment as a complete independent organism with its own excretory system, circular and longitudinal muscles to work the fluid-filled coelomic fluid as a **hydrostatic skeleton**, and a ganglion to coordinate events in the metamere. There were also paired **metanephridia** to filter coelomic fluid and remove metabolic wastes, as well as paired gonads on the septal wall of each metamere. As you work with your specimens and look through *Digital Zoology* be sure to make a note of metamerically arranged structures that you see both externally and internally.

✎ *Can you think of some of the advantages of having a body cavity?*

Closed circulatory system

With the metameric plan the advantage of the linear processing of ingested food made possible by a **complete gut** was now a liability. In other **eucoelomates** the coelomic fluid could be used to transport nutrients and wastes but in annelids the metameres are separated by **septa**. Coelomic fluid in adjacent segments doesn't mix between segments. It couldn't be used to transfer nutrients between metameres located next to parts of the gut that had become specialized for nutrient absorption and those next to regions where absorption did not occur. The development of the **closed circulatory system** in combination with metamerism was important in ensuring that nutrients were evenly distributed throughout the ancestral annelid.

With the coelomic fluid no longer having a major role in transporting nutrients, oxygen, and metabolic wastes, one of it's other important roles, as a hydraulic medium, took on new importance for the annelids.

✍ *What are some of the different functions of coelomic fluid?*

Burrowing - a new form of locomotion

In the ancestral annelid each metamere was hydrostatically isolated from the one next to it by the septal walls between the segments. Individual metameres could be lengthened, with the associated decrease in diameter, or shortened, with an increase in length, without affecting the shape of the adjacent segment.

All of this allowed annelids to be the first animals to burrow into the organic sediments of the ancient oceans–a new form of locomotion. But there was more to this than moving through the sediments. They didn't push the sediments aside as they burrowed–they swallowed them, becoming the first substrate feeders.

✍ *Substrate feeding is one of a number of different feeding strategies. What are some of the others?*

This ancestral characteristic of segmentation has been modified and only in the most primitive annelids is near perfect **serial homology** still visible down the length of the animal. In more advanced forms, regions of specialization have occurred and septa, which separated each segment, often disappear. In the leeches the body cavity is no longer as spacious and has once again become filled with tissue.

Increased activity associated with a new form of locomotion also made demands on other systems. It required changes in the nervous and sensory systems to coordinate movement. The circulatory system is more elaborate, as would be expected to supply the increased metabolic demands associated with this enhanced mobility and the necessity of transporting nutrients from the digestive tract to each of the isolated metameres. Increased activity increases metabolism, which in turn requires a more efficient means to deal with metabolic wastes. As a consequence the excretory system has become elaborate in annelids. Changes in the excretory system have also permitted the invasion of the terrestrial habitat but also set the limits for this phylum's success within this environment. Terrestrial worms have to absorb oxygen across their body walls.

A Closer Look Inside *Digital Zoology* – Annelids

Polychaeta: Nereid worm

The marine worm *Nereis* is often used to introduce the annelids because many of the ancestral annelid characters can be seen in the animal. But be careful as you make your observations. Remember that the ancestor to the annelids burrowed and there were no **parapodia**, something unique to the polychaetes. Bundles of **setal hairs** helped the ancestral annelid anchor its body in the burrow. Only later did lateral extensions of the body wall form parapodia, allowing the worm to move across the surface–the beginnings of the polychaetes and yet another way to get around. Take a close look at the parapodia and you'll see how they have their own musculature, oblique muscles, and strengthening **acicula**.

As the annelid moves across the substrate it needs to know where it's going, and the head has a variety of different sensory structures important for its predatory existence: eyes, jaws, and sensory tentacles are the most obvious. Internally each coelomic space is still isolated from the next and other than metanephridia, ganglia, nerve cord, and blood vessels, there really isn't much to see inside. Cross sections show how the muscles are arranged in four bands down the length of the segment, and this relates to how the animal moves. Be sure to look at the simple arrangement of the blood vessels, ventral nerve cord, and the simple tubular gut. Try and match the structures that you see in the photos of the cross sections with the pictures of the dissected specimens.

Although burrowing may not have been ancestral to the polychaetes, there are some worms in the class that do just that. Polychaetes are often grouped into two types: free swimming (errant) and sedentary. *Nereis* is an errant polychaete. For sedentary polychaetes the parapodia may get in the way, and they are often lost or modified for other functions. Two good examples of these changes are pumping water through the burrow and acting as gills. There also may be regional specialization down the length of the animal with parapodia doing different things depending on where they are found on the body. A great example is the parchment worm, *Chaetopterus*.

Clitellata: The earthworm

The earthworm gives us a chance to see the burrowing body plan that was ancestral to the phylum. Unlike polychaetes, which used muscles arranged in bands for movement, in earthworms they are arranged in layers, with an outer circular and inner longitudinal muscles. Depending on which is contracting, the whole segment is either lengthened or shortened. The numbers of setae are reduced and two pairs are found on each side of the animal. Cross section slides will show you these and their musculature.

As you look at the external and internal features of the earthworm, remember that this is a highly specialized animal that has conquered the terrestrial environment. Just as we saw in the polychaetes, there are now regions of specialization where the ancestral pattern of identical segments has disappeared. Externally the most obvious example of this is the clitellum. Internally the gonads are permanent and only found in a few of the anterior segments. The reproductive system is further complicated because these animals are monoecious (hermaphrodite) organisms. Each animal has structures to store its own sperm (seminal vesicle) prior to mating and needs a mechanism to pass it to another worm during mating, while simultaneously receiving the other's sperm and storing it (seminal receptacle) for later. The secret to monoecism is that the events of sperm transfer and fertilization are separate.

✍ *Why is it important that the events of sperm transfer and fertilization be separated in a monoecious organism?*

The circulatory system of the earthworm is more complex than that of the polychaetes, and five blood vessels that connect dorsal and ventral blood vessels have been enlarged and referred to as hearts. The surface area of the gut has been enlarged by an inner fold, the **typhlosole**, and in the anterior there are parts of the gut that have become specialized. The gut and many of the blood vessels, are covered in **clorogogen tissue**, that is involved in intermediary metabolism.

Clitellata: The leech

The number of segments in a leech is fixed, 32 to 34 depending on the species, but you wouldn't know it from looking at the external surface, which has extra rings, annuli, that hide the boundaries of the true segments. The anterior sucker with the mouth in its center, posterior sucker, male gonopore on segment 11, and the female opening on segment 12 are prominent external structures. The spacious coelomic cavities found in other annelids have disappeared in the leech, and all that remains are coelomic sinuses surrounding the internal organs. This return to a near-solid body plan makes a dissection difficult. In *Digital Zoology* prepared slides, *Glossiphonia sp.*, are used to examine the internal anatomy. These leeches have a proboscis that may be retracted or extended from the oral opening. The digestive tract is a straight tube with digestive ceca extending from the crop and intestine. The junction between the crop and intestine is identified by the last diverticula of the crop, which is larger than the others. Leeches are predatory, and the crop provides a location for storing food when not attached to a host. The multiple paired testes are connected to each other by the sperm duct and atrium that opens into the male gonopores. A single set of paired ovaries opens into the female genital pore. The ovaries and testis are referred to as ovisacs and testisacs because of the coelomic sinus that surrounds them. In cross sections you can see how the body cavity has been filled with longitudinal muscle and botryoidal tissue; remnants of the coelom are visible as sinuses and spaces surrounding the internal organs.

Cross-References to Other McGraw-Hill Zoology Titles

Integrated Principles of Zoology, 11th edition. C. P. Hickman, L. S. Roberts, and A. Larson. Chapter 17.

Animal Diversity, 3rd edition. C. P. Hickman, L. S. Roberts, and A. Larson. Chapter 11.

Zoology, 5th edition. S. A. Miller and J. P. Harley. Chapter 13.

Biology of the Invertebrates, 4th edition. J. Pechenik. Chapter 13.

Laboratory Studies in Integrated Principles of Zoology, 10th edition. C. P. Hickman, F. Hickman, and L. Kats. Chapter 12.

General Zoology Laboratory Guide, 13th edition. C. Lytle. Chapter 12.

Structures Checklist

Here are some of the structures that you should be able to easily find in *Digital Zoology* and the specimens that you will be looking at in your lab. After reading your lab handout, you might want to add more and, depending on the equipment available in your lab, you might see more. As you study the material, you might also want to make some notes on how some of these structures looked or include a drawing in your lab notes. (Structures indicated by * may be hard to see.)

Earthworm

External anatomy

- ☐ Anus
- ☐ Clitellum
- ☐ Female pores*
- ☐ Male pores
- ☐ Mouth
- ☐ Peristomium
- ☐ Prostomium
- ☐ Pygidium
- ☐ Seminal groove
- ☐ Setae
- ☐
- ☐

Internal anatomy

- ☐ Aortic arches (hearts)
- ☐ Calciferous gland
- ☐ Chloragogen tissue
- ☐ Crop
- ☐ Dorsal vessel
- ☐ Esophagus
- ☐ Gizzard
- ☐ Intestine
- ☐ Metanephridium
- ☐ Nerve cord
- ☐ Ovaries*
- ☐ Pharynx
- ☐ Pharynx musculature

- ☐ Seminal receptacles
- ☐ Seminal vesicles
- ☐ Septal walls
- ☐ Subneural vessel
- ☐ Testes*
- ☐ Ventral vessel
- ☐
- ☐
- ☐

Cross section

- ☐ Chlorogogen tissue
- ☐ Circular muscle
- ☐ Coelomic space
- ☐ Cuticle
- ☐ Dorsal vessel
- ☐ Giant axons

- ☐ Gut lumen
- ☐ Gut wall
- ☐ Longitudinal muscle
- ☐ Metanephridia
- ☐ Setae
- ☐ Setal musculature
- ☐ Subneural vessel
- ☐ Typhlosole
- ☐ Ventral nerve cord
- ☐ Ventral vessel
- ☐

Additional structures

- ☐
- ☐
- ☐

Nereid worm

External anatomy

- ☐ Anus
- ☐ Dorsal cirrus
- ☐ Eye
- ☐ Jaws*
- ☐ Mouth
- ☐ Neuropodium
- ☐ Notopodium
- ☐ Parapodia
- ☐ Peristomial cirri
- ☐ Peristomium
- ☐ Prostomial palp
- ☐ Prostomial tentacles
- ☐ Prostomium
- ☐ Pygidium
- ☐ Setae
- ☐ Ventral cirrus

- ☐
- ☐
- ☐

Internal anatomy

- ☐ Digestive caecum
- ☐ Dorsal vessel
- ☐ Intestine
- ☐ Jaws
- ☐ Longitudinal muscles
- ☐ Metanephridia
- ☐ Nerve cord
- ☐ Pharynx
- ☐ Septal walls
- ☐ Ventral vessel
- ☐

- ☐
- ☐
- ☐

Cross section

- ☐ Acciculum
- ☐ Coelomic space
- ☐ Dorsal vessel
- ☐ Gut lumen
- ☐ Gut wall
- ☐ Longitudinal muscles
- ☐ Neuropodium
- ☐ Notopodium
- ☐ Oblique muscles
- ☐ Setae
- ☐ Ventral nerve cord

- ☐ Ventral vessel
- ☐
- ☐

Additional structures

- ☐
- ☐
- ☐
- ☐
- ☐

Leech

Preserved specimen

- ☐ Annulations
- ☐ Anterior sucker
- ☐ Anus
- ☐ Eyes, if applicable
- ☐ Female genital pore*
- ☐ Male genital pore
- ☐ Mouth
- ☐ Nephridiopore*
- ☐ Penis, if inverted
- ☐ Posterior sucker
- ☐ Proboscis, if applicable
- ☐ Sensory papillae*
- ☐
- ☐
- ☐

Whole mount

- ☐ Anterior sucker
- ☐ Atrium (male)
- ☐ Brain
- ☐ Crop
- ☐ Crop diverticulum
- ☐ Eye
- ☐ Female gonopore
- ☐ Intestinal diverticulum
- ☐ Intestine
- ☐ Male gonopore
- ☐ Mouth
- ☐ Ovary (Ovisac)
- ☐ Posterior sucker
- ☐ Proboscis, if applicable
- ☐ Salivary gland

- ☐ Sperm duct (vas deferens)
- ☐ Testis (testisac)
- ☐
- ☐
- ☐

Cross section

- ☐ Botryoidal tissue
- ☐ Circular muscle
- ☐ Crop diverticulum
- ☐ Dermis
- ☐ Dorsal blood sinus
- ☐ Dorsoventral muscle
- ☐ Epidermal gland
- ☐ Epidermis
- ☐ Esophagus
- ☐ Gut lumen

- ☐ Gut wall
- ☐ Lateral blood sinus
- ☐ Longitudinal muscle
- ☐ Nephridium
- ☐ Oblique muscle
- ☐ Ovary (ovisac)
- ☐ Sperm duct (vas deferens)
- ☐ Testis (testisac)
- ☐ Ventral nerve cord and sinus
- ☐
- ☐
- ☐
- ☐

Crossword puzzle – Annelida

Puzzle solution. An interactive web-based version of this puzzle and its solution are available on the *Digital Zoology* website www.digitalzoology.com/students. With the interactive puzzle you can check to see if individual words or the whole puzzle is correct and get hints for single letters.

Across

1 These highly modified parapodia create and hold onto the mucus bag in the filter-feeding polychaete *Chaetopterus*. (7)

7 The specialized axons in the earthworm nervous system that help them escape into their burrows. (5)

9 Blood in the dorsal blood vessel of a worm moves toward this part of the body. (5)

12 The pygidium is at this end of an annelid. (9)

14 The extra rings of a leech. (6)

15 An earthworm's sperm is stored in the seminal _____ prior to mating. (8)

18 Compared to an oligochaete earthworm, marine polychaete worms have _____ setae. (4)

19 Metameres in a worm are also called these. (8)

20 The type of fluid you would find in both the metanephridia and coelomoducts. (8)

22 Deposit feeders eat this. (9)

26 These structures on the funnel of the metanephridia pump the coelomic fluid into the nephridia of an annelid. (5)

27 This structure increases the surface area of an earthworm's digestive system. (10)

29 The repeating units in an annelid have this type of homology. (6)

31 An earthworm has this number of paired setae on each body segment. (4)

32 This is the uppermost part of a marine polychaete worm's parapodia. (10)

33 Relative to the nerve cord, you'll find the ventral blood vessel in an earthworm here. (5)

35 The number of pairs of testes in an earthworm. (3)

37 Ingested calcareous rocks and stones are dissolved by this type of pH condition in the intestine of an earthworm. (6)

39 The number of hearts an earthworm has. (4)

40 Why there are three parts of an annelid that aren't segments can be found in this larval stage. (11)

42 Marine worms don't have these permanent organs; instead, they develop only during the reproductive season. (6)

43 This opening is found on the peristomium of a worm. (5)

Down

2 Each of the repeating segments in an annelid is referred to as one of these. (8)

3 In annelids, structures made from this embryonic tissue are not metamerically arranged. (8)

4 What you find inside the pharynx of the Nereid worm. (4)

5 The _____ esophageal ganglion is also called the brain (prefix). (5)

6 The supraesophageal ganglion of a worm is also called a this. (5)

8 This opening is found on the annelid's pygidium. (4)

10 In leeches, the coelomic cavity has been reduced to spaces referred to as these. (7)

11 Because the chlorogogue (chlorogogen) tissue is involved in intermediary metabolism, it is, in some ways, analogous to which human organ? (5)

13 These chitinous rods help support the parapodia of marine worms. (7)

16 In marine worms, the gonads form on these walls found between each metamere. (6)

17 Although this is the most anterior part of an annelid, it's not a true segment. (10)

21 In addition to the fluid from the metanephridia, coelomic fluid in annelids also reaches the outside through these. (12)

23 Parapodia are involved in locomotion and this function. (11)

24 The number of pairs of seminal vesicles in an earthworm. (5)

25 Where you would find the subesophageal ganglion relative to the esophagus in a worm. (5)

26 The circulatory system of an annelid is this type. (6)

28 These structures are important in both locomotion and gas exchange in marine worms. (9)

30 Marine worms, polychaetes, and earthworms, oligochaetes, have these but leeches, hirudinea, don't. (5)

31 The symbol for the metal found at the center of an oligochaetes respiratory pigment. (2)

32 Would you find hearts in the circulatory system of a marine worm? (2)

34 Free-swimming marine worms are referred to as being this. (6)

36 In annelids, these separate the metameres from each other. (5)

38 Any part of the digestive tract involved in storing food prior to digestion; it's particularly large in leeches. (4)

41 The symbol for the metal found at the center of a polycheates respiratory pigment. (2)

Self-Test - Annelida

Use the following labels to identify the photographs. You may have to use a label more than once, and some labels may not be appropriate for the photographs. Answers are available on the *Digital Zoology* website at www.digitalzoology.com/students. Be sure to try the interactive drag-and-drop quizzes on the *Digital Zoology* CD-ROM.

Specimens

A- Earthworm- *Lumbricus*

B - Leech

C- Nereid worm - *Nereis*

Labels

1) Brain
2) Circular muscle
3) Coelomic space
4) Crop
5) Dorsal vessel
6) Dorsoventral muscle
7) Epidermis
8) First seminal receptacle
9) Gizzard
10) Gut diverticulum
11) Gut lumen
12) Gut wall
13) Intestine
14) Longitudinal muscle
15) Metanephridium
16) Nerve cord
17) Oblique muscle
18) Pharynx
19) Second seminal receptacle
20) Second seminal vesicle
21) Septal wall
22) Sinus
23) Testisac (testis)
24) Third seminal vesicle
25) Typhlosole
26) Ventral vessel

MOLLUSCA

Inside *Digital Zoology*

As you explore the Mollusca on the *Digital Zoology* CD-ROM or on the web at www.digitalzoology.com, don't miss these learning tools:

- ○ Photos of dissceted specimens of the preserved clam, snail, and squid.
- ○ Drag-and-drop quizzes on the key structures of the clam, snail, and squid.
- ○ Interactive cladogram showing the major events that gave rise to the eight mollusc classes. A summary of the key characteristics of each class are combined with an interactive glossary of terms.

Defining Differences

The characteristics described in the following sections appear for the first time in the Mollusca, and they define the phylum. Others are important for understanding how animals in the phylum function. Molluscs have undergone adaptive radiation creating a bewildering array of animals that appear to have little or no resemblance to each other. It was once common practice in zoology courses to use a hypothetical mollusc that had the basic mollusc features to learn about animals in the phylum. It's unfortunate that it was called H.A.M.–hypothetical ancestral mollusc–because that gave the impression that this was what the ancestral mollusc might have looked like. It certainly didn't, and yet the exercise is a good one for understanding the phylum. So we'll rename H.A.M.–hypothetical aggregate mollusc–and still use it to discuss the different mollusc characteristics. You'll want to watch for examples of these in the various mollusc specimens that you will be examining.

- ○ Radula,
- ○ shell and mantle,
- ○ ctenidia,
- ○ ciliated visceral mass,
- ○ muscular foot, and an
- ○ open circulatory system.

Radula

If you've found an animal with a **radula**, you've found a mollusc. The radula gave molluscs a way of feeding completely different from how any other animal fed. If you think about the different feeding strategies that ancestors to various animal phyla had, you'll see why. Particulate suspension feeders sat on the substrate or floated in the ancient oceans to capture their food–sponges, echinoderms, and chordates. Flatworms glided across the ocean bottoms feeding on anything they could pick up with their fleshy muscular pharynx. That made platyhelminths one of the first to feed on the organic material that rained down and collected on the ocean bottom. Annelids tapped into a new unused food source by burrowing and consuming the substrate across which the flatworms were gliding. While all this was going on, predatory cnidarians were feeding on other animals in the zooplankton.

What about the organic matter "stuck" to the substrate that a flatworm couldn't get with only a fleshy pharynx? The solution was to scrape it off, and molluscs used their radula to do that. It's a simplistic view of things but for any animals that can exploit an unused food source there is the potential to diversify into new species. For molluscs, the radula, combined with the other traits that you will be looking at, allowed this phylum to undergo the extensive **adaptive radiation** that has made molluscs the second largest animal phylum. (third, if there are as many nematodes as biologists think).

Most textbooks compare the radula to a file, scraper, or rasp in a wood worker's tool box. (If you're more the power tool type–a belt sander.) It's a good analogy; the radula is a ribbon of teeth supported underneath by the tonguelike **odontophore**. As the radula is extended and retracted, it scrapes away at the substrate. If you rubbed a file on a piece of wood you would generate wood shavings or sawdust. It is the same with a radula, which creates particulate food. The usual solution for getting particulate food into the digestive tract is to trap it in **mucus**, then use cilia to propel the food-laden string of mucus into the mouth. As you might have guessed, that's what molluscs do. As the radula grinds away at the substrate the teeth on the end are worn down. They're replaced by new teeth added at the back of the radula, which as it grows, moves forward on the surface of the odontophore.

✍ *How does a mollusc separate the organic food from the inorganic substrate that it ingests?*

Shell and mantle

The shell originated as a defensive strategy by early molluscs to protect themselves using calcareous **spicules**, or spines, embedded in their outer epidermis. Spicules can still be found along with edges of some chiton shells and in some of the unshelled molluscs. Over time, the spicules became larger, forming plates that later fused with each other to create a solid shell covering the entire dorsal surface of the animal. The specialized epidermis and its gland cells that produced the shell are the **mantle**. **Retractor muscles** attached to inner surface of the shell extended into the foot on the ventral side and when these muscles contracted the shell was pulled down against the substrate. The mollusc hid inside waiting for whatever danger lurked outside to go away.

✍ *What are the three layers of the mollusc shell and of what is each made?*

It was a good defensive strategy but while it solved one problem, it created another. The new shell decreased the available surface area for gas exchange and any other functions that depended on diffusion. The problem was solved by extending the shell over the edge of the body, creating an open space underneath it–the **mantle cavity**. Just as the mantle on a fireplace hangs over the cavity of the fireplace, that's what the mollusc's mantle does, and the increased surface area for gas exchange occurs as gills that developed inside the cavity.

Ctenidia

Mollusc gills are called ctenidia, and when the cilia that cover them beat, water is pulled into the mantle cavity, across the surface of the ctenidia, and out the cavity. The ctenidia are hollow and the spaces inside of them are part of the hemocoel. As blood flows through ctenidia, it is oxygenated by the water passing by the outer surface.

The path that water takes through the mantle cavity is unidirectional. Once the water is past the ctenidia and on its way out, it goes by the openings of the **metanephridia**, gonad, and anus picking up gametes, if it's the reproductive season; metabolic wastes; and any undigested food from the alimentary tract, carrying it away from the mollusc.

Ciliated visceral mass

As important as cilia are in how the ctenidia work, they are equally important in most of the internal systems of a mollusc and why we make the point about a **ciliated visceral mass**. Cilia will propel the food through the digestive tract, and cilia on the metanephridia help propel the coelomic fluid into the excretory apparatus. Almost anything a mollusc does is done by using cilia, with one notable exception–how it moves.

Muscular foot

Although cilia initially played a role in movement, and may still be found on the bottoms of some mollusc feet, muscles in the foot have become more important for how molluscs move; it's the reason the muscular foot is a molluscan trait.

There are four things that we need to keep track of to be able understand how the mollusc foot works, two of these are muscles and the other two are not muscles. One of two sets of muscles extend from the shell into the foot along the dorsal ventral axis of the body–the **dorsoventral muscles**. The other set of muscles run at right angles to the dorsoventral muscles from one side of the foot to the other–the **transverse muscles**. The two sets of muscles work in combination with the blood in the **hemocoel**, the third part, to form a **hydrostatic skeleton**. Think of the foot as a fluid-filled box. (When we looked at a hydrostatic skeleton in cnidarians we had sheets of muscles and a fluid-filled balloon. This is different.) If we contract the dorsoventral muscles, the box could increase in either length, width, or both. But when the dorsoventral muscles contract, the transverse muscles contract at the same time. The result is that the box can only increase in length.

At the anterior end of the foot, contraction of the muscles extends the foot forward, and its tip contacts the substrate farther in front of the animal. Because the dorsoventral muscles have contracted, the part of the foot just behind where the most anterior is now touching, is raised off the substrate. Here's where the fourth element, **mucus**, comes in. The edges of the raised space underneath the contracted dorsoventral muscles are sealed to the substrate by mucus, like a small suction cup. As the dorsoventral muscles at the back of the suction cup contract, lifting that end, the dorsoventral muscles at the front relax. Because the space is sealed by mucus as the back end is lifted, the front end is sucked down, stretching the previously contracted dorsoventral muscles. A wave of contractions moves toward the back of the animal as the animal moves forward, a **retrograde wave**.

In some molluscs, each side of the foot is capable of separate waves of contraction, and depending on how fast the contractions move down either side, the animal turns and changes directions. It's like the treads on an armored tank, and with the way a mollusc protects itself using the shell, the analogy is even more appropriate.

Open circulatory system

In most zoology courses the molluscs are the first animals you'll encounter with an **open circulatory system**. In a closed circulatory system, exchange between the circulating blood and various tissues and organs occurs across **capillaries**. They're missing in an open circulatory system. Instead, blood pools in spaces or cavities, the hemocoel, and bathes the organs and tissues.

The hemocoel is the main body cavity of a mollusc and an important part of how molluscs move. There is a true coelom, but it is small, surrounding the heart. That's how it gets its name, the **pericardial cavity**. Gonads are found on the wall of the pericardial cavity, and an open funnel-shaped metanephridia filters the coelomic fluid in the cavity to remove metabolic wastes.

 How do metabolic wastes get from blood in the hemocoel to the fluid in the pericardial cavity that the metanephridia filters?

A Closer Look Inside *Digital Zoology* - Molluscs

Bivalvia: Freshwater clam

Clams and their bivalve relatives are one of only a few mollusc classes found in both freshwater and marine environments. In both places they're specialized as filter feeders, and like most filter feeders they're essentially sessile. To accommodate their way of life, bivalves have made some major modifications to the mollusc body plan. The most obvious is that the body is compressed laterally with the mantle folded down over the sides in two lobelike sheets. The shell matches the shape of the mantle's two parts. The shell has two valves that cover each side of the animal and the sides are connected by an elastic **hinge ligament** across the top. The two valves give the class their name, the bivalvia. But, be careful. These are **univalve** molluscs because the original single valve is only folded! Another modification for the filter-feeding lifestyle has resulted in the loss of the radula and cephalization.

 How does a clam burrow in the sediment?

If you take a look at the **cladogram** for the molluscs in *Digital Zoology* or your textbook, you'll see that bivalves are more advanced than cephalopods, which include squids and octopods. Most students have trouble understanding how something like a highly intelligent octopus is considered more primitive than a sessile filter-feeding clam! **Cladistics** looks for shared characters and how they are modified from the ancestral form. If the character disappears completely, that's considered advanced. The things we think make the squids and octopods superior animals aren't characteristics that define the phylum. A radula defines the molluscs, and cephalopods still have their radula, and bivalves don't. It's one of a variety of reasons why the two groups are placed where they are on the cladogram.

The oldest part of the clam's shell is the **umbo,** and as the animal grows, the edge of the mantle grows out from the umbo. A clam keeps its shell closed by the contraction of large anterior and posterior **shell adductor muscles.** When the two muscles relax, the elasticity of the hinge opens the shell. It's these adductor muscles that let you know whether you should eat a steamed clam or mussel. Steaming kills the clam, and in death, muscles that hold the shell closed, relax and the shell pops open. If the shell didn't pop open when you cooked the clam, it was dead before it made it to the steamer.

✍ *What causes the ridged rings on the surface of the clam's shell?*

In the lab, once you've taken the shell off your clam you'll see the mother-of-pearl layer, or **nacre**, on the inside of the shell, along with the **pallial line** where the mantle connects and the **muscle scars** where the muscles attach. Fold the mantle back into place over the foot and ctenidia so that you can see how the left and right sides form the **incurrent siphon** and **excurrent siphon**. Freshwater clams have large ctenidia, which might lead you to believe that the animals need a lot of oxygen for something. However, they don't. The ctenidia are primarily involved in feeding. Cilia on the surface of the ctenidia pull water into the mantle cavity and filter out particulate food, passed forward along the edge of the ctenidia, to the **labial palps** and into the **mouth**.

✍ *Describe the route that water follows as it passes through a clam starting from the incurrent siphon and finishing with the excurrent siphon.*

The lateral compression of the bivalve body not only changes the shape of the shell but also the organs of the visceral mass. Instead of sitting above the muscular foot, the visceral mass is partially buried in the foot. The only part still visible above the foot is the pericardial cavity on the dorsal side. Inside it you'll see the **heart** that wraps around the **intestine** that passes through the cavity. If you are careful you'll also see the **atrium** to the heart. If you can't see it in your specimen, take a look in *Digital Zoology*. The atrium is important, and if you answered the question on how a metanephridia filters metabolic wastes, you'll know why. If you didn't answer that question earlier, now would be a good time.

To see the other organ systems in the foot, you'll have to patiently carve slices off the foot and slowly work your way deeper into the foot. As you do, the intestines, **stomach**, **digestive glands**, and **gonads** will appear surrounded by the spongy musculature of the foot. Those spaces that create the sponginess are part of the hemocoel. If you have trouble seeing any of these, there are pictures in *Digital Zoology*.

Gastropoda: The snail

The gastropod trait of **torsion** makes the snail one of the more difficult invertebrate dissections. The **visceral mass** on top of the **muscular foot** has been rotated 180 degrees and the **mantle cavity** is now in front, sitting above the head. The opening of the mantle cavity points forward instead of toward the back, as it originally did. Another result of torsion is that structures caught on the inside of the bend are lost.

Snails often only have one of what would have been paired organs. These molluscs are **asymmetric** rather than bilaterally symmetric.

 What are the presumed advantages of torsion, and for what stage in the gastropod life cycle was it an advantage?

To make things more complicated, snails are monecious, and to save space inside the cramped shell the two reproductive systems share some of same ducts. If that isn't confusing enough, the visceral mass is coiled inside a spiral shell! Be careful, though. The spiral shell is not a direct consequence of torsion. It's a solution for packaging the visceral mass. As the snail grows, the size of the visceral mass on top grows as well. It's wound up into the shell.

The snail's shell is an ideal hiding place for protection against predators, but it also protects against dissection. Snails are found in terrestrial, as well as freshwater and marine, environments. Terrestrial snails have lungs and breathe by raising and lowering the floor of the mantle cavity. This pumps air in and out of the **pneumostome** located in the **collar** between the shell and foot. The successful invasion of land accounts for the success of the class.

Once you have your snail out of its shell, you'll have to start unwinding the viscera. Because of torsion, functional systems in the animal will wind their way up to the tip of the whorl and back down. The best example of this is the digestive system, which starts with the **buccal mass** under the mantle cavity, followed by the esophagus and the **stomach,** located in the first part of the whorl. The **intestine** leaves the whorl, passing by the first parts of the digestive tract on its way the anal opening. The digestive gland extends up and to the tip of the whorl.

The reproductive tract does the same thing, and some parts of the reproductive tract are easier to see than others. Use the pictures in *Digital Zoology* to help you orient with some of the more obvious landmarks in your specimen. The **dart sac**, **mucus gland**, **albumin gland**, and **flagellum** are usually the easiest to see. Once you've found those you should be able to follow the various tubes to find the **penis**; combined vas deferens and **oviduct**; **hermaphrodite duct**; seminal receptacle; and if you were especially careful, the **ovotestis** buried in the digestive gland.

 In snails, describe the events from insemination to oviposition.

Cephalopoda: The squid

A cephalopod's visceral mass has been stretched along the dorsoventral axis above the foot, bringing the head and foot closer together on the ventral side. That's how they got their name, Cephalopoda (head, foot). The mantle surrounds the visceral mass, and ancestrally a hard **shell** surrounded all of this to form an elongated cone-shaped shell with the head and foot poking out the open end. It was easier to point the tip of the shell in the direction that the animal was moving. Cephalopods swim with their dorsal surface pointing forward, rather than up.

With a larger visceral mass on top of a muscular foot, cephalopods were faced with the same problem as gastropods had for finding a way to make a more compact body. They used the same solution, although they wound their shells in a different way.

The fossil record includes cephalopods with unwound, wound, and even partially wound shells, and the diversity of fully wound ammonite fossils is good evidence of the successful body plan. At one time these cephalopod predators ruled the ancient oceans, but now only *Nautilus* remains to give a hint to what these ancient animals looked like.

✍ *How does the nautalid shell differ from that of a gastropod?*

Shelled cephalopods disappeared at about the same time as jawed predatory fishes appeared. It's possible that the two events are related. The mollusc adaptation to these more active and agile newcomers was another modification to the mollusc body plan to become just as active and agile–the result are the cephalopods we see today.

To live this new active and predatory life there have been extensive changes in how a cephalopod, like the squid, functions compared to other molluscs. Watch for these changes as you work with your squid in the lab or as you look at *Digital Zoology*.

One of the most obvious changes is the loss of the shell, or if it's still present a major reduction in its size. Externally you can't see a shell in your squid until you open the mantle and see the **pen** inside. Without a shell, the mantle is much more flexible, and its musculature is used to pump water in and out of the mantle cavity and across the gills. In other molluscs, cilia were responsible for this but they are completely missing on the gills of cephalopod. Take a look at the edge of the mantle on your squid, and you'll see the **mantle cartilages** that work like a lock and key to hold the mantle edge to the body. The mantle expands, pulling in around its edge. When it contracts, water is forced out the **funnel**. This movement of water doesn't only aerate the gills, it jet propels the animal. It's a new form of locomotion.

This enhanced water flow across the gills supplies levels of oxygen high enough to meet the metabolic demands of being an active predator, even though its flow is concurrent with the blood inside the gills. An open circulatory system with a hemocoel is not the most efficient way to move blood. Another cephalopod improvement is a closed circulatory system with capillaries supplying blood to the organs and tissues. To increase the efficiency of the system, a pair of **branchial hearts** push the blood across the gills to the systemic heart, which then pumps it out to the rest of the body. The branchial hearts are next to the metanephridia to remove metabolic wastes.

✍ *What route does blood follow as it moves through the squid? Start at the systemic heart.*

Being an active predator requires an advanced sensory and nervous system to locate and trap prey and to locate a mate and be able to differentiate that mate from a potential meal, or for that matter becoming one. Components of the system include a well-developed **brain** enclosed in a **cartilaginous brain case** and eyes analogous to our own. Cephalopods use **elastic capsule chromatophores** to change colors to signal mates or as camouflage. There are some cephalopods that can produce light, and their blinks and flashes are a communications tool.

Even the digestive tract has been extensively modified to fuel the metabolic demands. The **muscular foot** now forms the eight **arms** and paired **tentacles** used to capture prey using the suction-cuplike **suckers**. There is still a **radula**, but a parrotlike **beak** is at the center of the ring of tentacles and arms, and it rips at food, which is swallowed and passed back into a digestive system. A **digestive caecum**, which acts like a crop as well, is the site of nutrient absorption. Other parts of the digestive system include the **stomach** and **digestive gland** (often called a liver), which produces digestive enzymes, and a small pouch on the digestive gland, the **pancreas**, which also produces digestive enzymes. The **esophagus** is buried inside the digestive gland and after the digestive caecum the digestive tract bends back on itself as the intestine leads toward the anus. Undigested food is dumped into the excurrent flow of the mantle cavity.

 How do cephalopods solve the problem of getting sperm from the male to the genital pore of the female, hidden inside the mantle cavity?

Cross-References to Other McGraw-Hill Zoology Titles

Integrated Principles of Zoology, 11th edition. C. P. Hickman, L. S. Roberts, and A. Larson. Chapter 16.

Animal Diversity, 3rd edition. C. P. Hickman, L. S. Roberts, and A. Larson. Chapter 10.

Zoology, 5th edition. S. A. Miller and J. P. Harley. Chapter 12.

Biology of the Invertebrates, 4th edition. J. Pechenik. Chapter 12.

Laboratory Studies in Integrated Principles of Zoology, 10th edition. C. P. Hickman, F. Hickman, and L. Kats. Chapter 11.

General Zoology Laboratory Guide, 13th edition. C. Lytle. Chapter 11.

Structures Checklist

Here are some of the structures that you should be able to easily find in *Digital Zoology* and the specimens that you will be looking at in your lab. After reading your lab handout, you might want to add more and, depending on the equipment available in your lab, you might see more. As you study the material, you might also want to make some notes on how some of these structures looked or include a drawing in your lab notes. (Structures indicated by * may be hard to see.)

Clam

External anatomy

- ☐ Anterior end
- ☐ Excurrent siphon
- ☐ Growth lines
- ☐ Hinge ligament
- ☐ Incurrent siphon
- ☐ Left valve
- ☐ Periostracum
- ☐ Posterior end
- ☐ Right valve
- ☐ Umbo
- ☐
- ☐
- ☐
- ☐
- ☐

Internal anatomy

- ☐ Anterior adductor muscle and scar
- ☐ Anterior aorta
- ☐ Anterior foot retractor muscle and scar
- ☐ Atrium
- ☐ Crystalline style*
- ☐ Ctenidia
- ☐ Digestive gland
- ☐ Esophagus
- ☐ Food groove on gill lamella
- ☐ Foot
- ☐ Gonad
- ☐ Heart
- ☐ Hinge teeth
- ☐ Intestine in foot
- ☐ Intestine in pericardial cavity
- ☐ Labial palp
- ☐ Mantle
- ☐ Metanephridium
- ☐ Mouth*
- ☐ Pallial line
- ☐ Pericardial cavity
- ☐ Posterior adductor muscle and scar
- ☐ Posterior aorta
- ☐ Posterior foot retractor muscle and scar
- ☐ Stomach
- ☐ Suprabranchial chamber
- ☐
- ☐
- ☐
- ☐
- ☐
- ☐

Additional structures

- ☐ Glochidia
- ☐
- ☐
- ☐
- ☐

Snail

External anatomy

- ☐ Anus*
- ☐ Coiled visceral mass
- ☐ Collar
- ☐ Eyes on second tentacle*
- ☐ Foot
- ☐ Genital pore*
- ☐ Growth lines
- ☐ Head
- ☐ Mantle
- ☐ Mouth
- ☐ Pneumostome
- ☐ Shell aperture
- ☐ Shell apex
- ☐ Tentacles
- ☐ Whorl
- ☐
- ☐
- ☐
- ☐
- ☐

Internal anatomy

- ☐ Albumin gland
- ☐ Buccal mass
- ☐ Crop
- ☐ Dart sac
- ☐ Digestive glands
- ☐ Esophagus
- ☐ Flagellum
- ☐ Hermaphrodite duct
- ☐ Intestine
- ☐ Lung
- ☐ Mantle cavity
- ☐ Mucus gland
- ☐ Oviduct
- ☐ Ovitestis
- ☐ Penis
- ☐ Salivary glands
- ☐ Seminal duct
- ☐ Seminal receptacle
- ☐ Sperm duct (vas deferens)
- ☐ Stomach*
- ☐
- ☐
- ☐
- ☐
- ☐
- ☐

Additional structures

- ☐ Shell columella
- ☐
- ☐

Squid

External anatomy

- [ ] Anterior surface
- [ ] Arms
- [ ] Beak
- [ ] Buccal membrane
- [ ] Collar
- [ ] Dorsal surface
- [ ] Eyes
- [ ] Head
- [ ] Lateral fins
- [ ] Mantle
- [ ] Mantle cartilages
- [ ] Mantle collar
- [ ] Posterior surface
- [ ] Suckers
- [ ] Tentacles
- [ ] Ventral surface
- [ ]
- [ ]
- [ ]
- [ ]

- [ ]
- [ ]

Internal anatomy

- [ ] Afferent branchial vein
- [ ] Anterior vena cava (cephalic vein)
- [ ] Brain
- [ ] Branchial hearts
- [ ] Cephalic aorta
- [ ] Cephalic cartilage
- [ ] Digestive caecum
- [ ] Digestive gland (liver)
- [ ] Efferent branchial vein
- [ ] Esophagus*
- [ ] Funnel retractor muscles
- [ ] Gills

- [ ] Head retractor muscles
- [ ] Ink sac
- [ ] Intestine
- [ ] Mantle arteries
- [ ] Metanephridia (kidneys)
- [ ] Nidamental gland
- [ ] Ovaries
- [ ] Oviducal opening
- [ ] Oviduct
- [ ] Pancreas
- [ ] Pen
- [ ] Posterior aorta
- [ ] Posterior vena cava (lateral abdominal vein)
- [ ] Rectum
- [ ] Spermatophoric duct
- [ ] Spermatophoric sac (Needham's sac)

- [ ] Systemic hearts
- [ ] Testes
- [ ]
- [ ]
- [ ]
- [ ]
- [ ]
- [ ]

Additional structures

- [ ] Hectocotylized arm in male
- [ ]
- [ ]
- [ ]
- [ ]
- [ ]
- [ ]
- [ ]

Crossword Puzzle – Mollusca

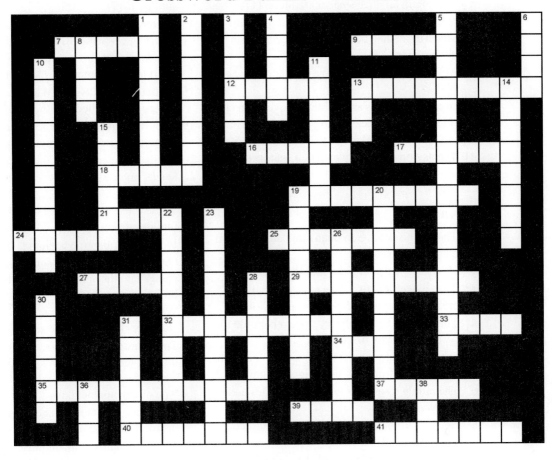

Puzzle solution. An interactive web-based version of this puzzle and its solution are available on the *Digital Zoology* website www.digitalzoology.com/students. With the interactive puzzle you can check to see if individual words or the whole puzzle is correct and get hints for single letters.

Across

7 Describes each half of a clam's shell. (5)

9 Relative to the other layers of a mollusc shell, the position of the periostracum. (5)

12 The number of major layers that make up a mollusc shell. (5)

13 In clams, the heart wraps around this as it passes through the pericardial cavity. (9)

16 This ligament makes the clam shell pop open when the adductor muscles relax. (5)

17 The bivalves, like clams, are specialists at this type of feeding. (6)

18 One way to describe the molluscan body plan is a visceral mass that relies on these to function on top of a muscular foot. (5)

19 The inner shell layer in a mollusc. (9)

21 The oldest part of the bivalve clam shell. (4)

24 The blood passes through these just before entering the heart in a mollusc. (5)

25 The ctenidia are extra big in clams because they are specialized for doing this. (7)

27 There are thousands of these on the surface of the radula, and they can scrape, pierce, tear, or cut at a mollusc's food. (5)

29 Instead of ancestral, the A in H.A.M. should be this. (9)

32 The eye in a squid is this compared to the human eye. (9)

33 Terrestrial snails breath using this instead of gills. (4)

34 The shell of a clam is composed of this number of valves. (3)

35 These parts of a closed circulatory system missing in an open system are these. (11)

37 You can tell where the muscles of the clam attach to its shell because you'll see these on the shell surface. (5)

39 Molluscs have this type of a circulatory system. (4)

40 In squids the gills are no longer covered in cilia now that these are used to pump water in and out of the mantle cavity. (7)

41 The asymmetric body plan of a snail is a result of this. (7)

Down

1 The molluscs have an open circulatory system and blood pools here. (8)

2 Describes the molluscan gill. (8)

3 The edge of this tissue secretes the shell and creates the cavity that bears its name. (6)

4 This part of the circulatory system is inside the pericadial cavity of molluscs. (5)

5 Digestive enzymes in the stomach of the clam are released from this spinning structure. (11,5)

6 The radula resembles this type of tool found in a toolbox. (4)

8 Because of torsion, you'll find this on top of a snail's head. (4)

10 As a result of torsion, snails have this kind of symmetry. (10)

11 One of the other larval stages common in molluscs. (7)

13 Secretions from this gland act as a decoy allowing cephalopods to make their escape. (3)

14 The more technical term for the mother-of-pearl layer of a mollusc shell is this layer. (8)

15 Particulate food is embedded in a string of this before being pulled into the clam's stomach. (6)

19 Squids and octopods are molluscs adapted to this type of feeding lifestyle. (9)

20 Snails are referred to as this because both sexual organ systems are found in each animal. (10)

22 These sense organs in the mantle cavity of a mollusc detect water quality and condition. (9)

23 In molluscs, the true coelom is this cavity. (11)

26 Once the food has been sorted in a mollusc, it passes into this gland to be biochemically broken down. (9)

28 When the shell adductor muscle contracts in a clam, this happens to the shell. (6)

30 There's no ciliated larval stages in cephalopods; instead, miniature animals that resemble the adults hatch from the egg. It's referred to as this type of development. (6)

31 Unlike the flatworms that came before them, the molluscs added this to their mesoderm. (6)

36 All that remains of the mollusc shell in a squid is a short rod referred to as this. (3)

38 The material that flows through a pneumostome. (3)

Self-Test - Mollusca

Use the following labels to identify the photographs. You may have to use a label more than once, and some labels may not be appropriate for the photographs. Answers are available on the *Digital Zoology* website at www.digitalzoology.com/students. Be sure to try the interactive drag-and-drop quizzes on the *Digital Zoology* CD-ROM.

Specimens

A- Clam
B- Snail
C- Squid

Labels

1) Afferent branchial vein
2) Albumen gland
3) Anterior shell adductor
4) Anterior vena cava
5) Branchial heart
6) Ctenidia
7) Dart sac
8) Digestive caecum
9) Digestive gland
10) Excurrent siphon
11) Flagellum
12) Foot
13) Heart
14) Hermaphrodite duct
15) Hinge
16) Incurrent siphon
17) Intestine
18) Mantle
19) Mantle cartilage
20) Mouth
21) mucus gland
22) Oviduct
23) Pancreas
24) Penis
25) Pericardial cavity
26) Posterior shell adductor
27) Seminal receptacle
28) Stomach
29) Suprabranchial chamber
30) Vas deferens

NEMATODA; GASTROTRICHA; ROTIFERA, INCLUDING ACANTHOCEPHALA

Inside *Digital Zoology*

As you explore the Nematoda, Gastrotricha, and RotiferA on the *Digital Zoology* CD-ROM or on the web at www.digitalzoology.com, don't miss these learning tools:

❍ Photos of dissected specimens and cross sections of the pig roundworm, *Ascaris lumbricoides*; prepared slides and photos of gastrotrichs; the rotifers, *Asplanchna* and *Philodina*; and male and female acanthocephalans.

❍ Drag-and-drop quizzes on the key structures of *Ascaris*, gastrotrichs, rotifers, and acanthocephalans.

❍ Video clips showing the whiplike movement of *Turbatrix aceti* (the vinegar eel) and movement, ciliary feeding, and the mastax in rotifers.

❍ Interactive cladograms showing the major events that gave rise to this unusual group of animals often described as pseudocoelomates or aschelminthes. Watch for differences on where they appear in the cladograms for molecular and traditional morphological interpretations. A summary of the key characteristics of each taxon are combined with an interactive glossary of terms.

Should there be a phylum **Aschelminthes** that includes seven or eight taxa linked by **monophyletic** characters? Or are these taxa an artificial group, the **pseudocoelomates**, with nothing to link them together but a few **convergent traits**? The debate rages in zoological circles, and the taxonomic status of these animals, which only a few years ago were collectively referred to as the pseudocoelomates, is being revised once again. Things have gone full circle from when the Aschelminthes were a phylum, to the term being dropped because the taxa in the phylum weren't related to each other. Now some biologists feel that these animals are linked by the presence of a cuticle and the absence of **cilia** or **microvilli** on their **ectodermal** surfaces and want to bring back the phylum Aschelminthes.

There's more involved in the debate than the validity of the Aschelminthes, and some scientists feel that molecular evidence requires even further revisions of the classification and that a whole new taxon, the **Ecdysozoa**, be added. This would include the old pseudocoelomates, calling them again the Aschelminthes, and the arthropods. Monophyletic characters uniting the group are that all these animals molt and lack ectodermal cilia. Both points of view are accommodated in *Digital Zoology* with cladograms based on molecular data and traditional morphological characters. This chapter of the workbook includes some of the animals caught in the controversy.

Defining Differences

Some of the differences described in the following sections appear for the first time in the different phyla, and they define the phylum. Others are important for understanding how animals in the phylum function. Whichever is the case, you'll want to watch for examples of these in the specimens that you will be examining.

Nematoda

❍ A pseudocoel,
❍ longitudinal muscles
❍ collagenous cuticle,

- ○ sensory amphids, and a
- ○ complete gut.

Gastrotricha

- ○ Presence of cuticle and cilia,
- ○ eutely, and
- ○ sticky adhesive tubes to anchor to the substrate.

Rotifera

- ○ Mastax and ciliated corona used for feeding,
- ○ reproduce by parthenogenesis, and
- ○ demonstrate cryptobiosis.
  Parasitic forms (Acanthocephala) with
- ○ unique ligaments, ligament sacs, and a uterine bell
- ○ and a spiny proboscis.

Nematoda

Pseudocoel

Unlike a true **coelom**, a **pseudocoel** isn't completely lined with **mesoderm**, and a **peritoneum** doesn't completely surround the cavity. Instead, the mesoderm of a pseudocoel is found adjacent to only structures formed from **ectoderm** and not those derived from **endoderm**. A consequence of this is pseudocoelomate animals don't have any musculature lining the gut, nor do they have mesenteries holding the internal organs in place.

 How do pseudocoelomates swallow their food?

One of the problems of using the pseudocoel as a unifying character for all these animals is that not all of these animals have body cavities. In some specimens, what was once thought to have been a pseudocoelomic space was an artifact from how the samples had been prepared or the specimens preserved, and these animals are actually **acoelomate**. At one point biologists thought they had it figured out. They thought the pseudocoel was the remains of the **blastocoel** that persisted in the adult. In some animals it remained as an open space, and in others it filled in again. But, even that is open to debate. One thing there is no debate about is that nematodes have a body cavity and that it fits the traditional definition of a pseudocoel.

Longitudinal muscles

Nematodes have only longitudinal muscles, and with a single fluid-filled pseudocoel acting as a **hydrostatic skeleton**, any contraction of the musculature alters the shape of the whole animal. It's the reason why movement in nematodes is characteristically whiplike, with wavelike undulations running the length of the body. One of the unusual features of nematode muscles is that it's the muscles that send cytoplasmic arms to connect with the ventral or dorsal nerve cords. That's the complete opposite of what you see in other animals where the nerves extend cytoplasmic extensions, axons, to the muscles.

Collagenous cuticle

Nematodes, like a variety of other animals, have a nonliving outer covering, the **cuticle**, produced by the epidermis underneath. In most invertebrates that don't have a cuticle, the outer epidermal surface is covered in cilia, which have a variety of functions. The presence of an outer cuticle changes that, and not surprisingly, wrapping a nonliving outer covering over the epidermal surface means cilia can't function. They have disappeared from the surface of animals with cuticles.

Invertebrates have cuticles made of either **collagen**, protein polymers, or **chitin**, carbohydrate polymers. A nematode's cuticle is collagenous and made up of three layers that form a mesh, or lattice, of fibers. The collagen fibers are inflexible but, the way that they're organized in the mesh allows the body of the nematode to bend and flex. Think of the protective gate placed at the top of a flight of stairs to keep small children from accidentally falling down the stairs. The wooden bars in the gates are inflexible, like the collagen fibers, but the gate can change its shape becoming long and narrow or short and high. That's the secret to the hydrostatic skeleton and cuticular covering of nematodes.

Sensory amphids

Nematodes have a pair of unique sensory structures at the anterior end of the body called **amphids**. They open to the outside through a pore, and an amphid pouch is found at the base of a short duct. These are sensory structures and a nerve connects them to the brain that surrounds the **pharynx**. Inside, the amphids are ciliated, and it's the only place that you'll find cilia in the nematode. Until cilia were found here, it was assumed that nematodes had completely lost their cilia.

Complete gut

Nematodes certainly aren't the first animals to have a **complete digestive tract** that includes a mouth at the anterior end and anus at the other end. But in zoology books it's common practice to discuss its advantages once we get to the first animals that have one, and that's why we're doing it here with the nematodes that have the complete gut with regional specialization. Having this type of digestive system may not seem like a big thing but it was, and here's why.

With an **incomplete digestive tract**, any undigested food must be removed from the system by the opening used for getting new food into it, the mouth. Imagine an animal with a partially digested meal in an incomplete gut when it encounters more food. There are two ways to deal with this, either remove the old food from the gut and then consume the new meal or consume a smaller meal and add it to what's already in the digestive system.

It's a problem that animals with a complete gut don't have. They can feed whenever food is available. More food can be added at the anterior end, as partially digested food is moved back toward the posterior end of the animal and ultimately out the anus. Food can now be processed in a linear sequence, and whenever an opportunity to feed occurs, it can happen. This type of organization is also described as a tube-within-a-tube digestive system, and it allows for regions of specialization to occur in the digestive tract. These can include ways to grind and mix the food and specialized regions where different digestive conditions, such as acidity or alkalinity, can be varied.

Another advantage of this type of construction is that the gut can move independently of the body wall. An animal can appear to be sessile while peristaltic movements of the gut musculature assist with digestion of ingested food. Nematodes and pseudocoelomates don't have mesoderm adjacent to their endoderm, and musculature doesn't line most of their gut. In these animals, body movements are still important for propelling food through the alimentary.

Gastrotricha

Cuticle and cilia

Gastrotrichs are unusual in having both cilia on the ventral surface and cuticular plates on the remainder of the body. What makes this unusual is that cilia are also covered with a fine layer of flexible cuticle. Many biologists believe that the gastrotrichs are some form of intermediate between ciliated animals and those covered in cuticle. The intermediate nature of the phyla was strengthened by what was thought to be a pseudocoel, an intermediate condition between acoelomate and eucoelomates. The position of the gastrotrichs is still not clear, and if you look at the cladograms in *Digital Zoology* you'll see just how ambiguous their place is in the animal evolution.

Eutely

Eutely is common in minute or microscopic animals, and it may be the result of natural selection for miniaturization. Too often we think of animals evolving and becoming bigger and perhaps better. There is an alternate strategy where some animals become smaller inhabiting microniches unavailable to other organisms; the water between grains of sand is a good example. Presumably as animals became smaller, there was a lower limit to the number of cells that any organ or the animal could have. Any fewer was detrimental and resulted in death; any more, a disadvantage. This may explain why many of the animals in the **meiofauna**, including gastrotrichs, have constant numbers of nuclei or cells for either organ systems or in the whole animal. Species can be distinguished from each other by the number of nuclei.

Adhesive tubes

The tips of each end of the forked tail of the gastrotrich contain adhesive glands that secrete cement that anchors the gastrotrich in place. The system uses a **dual gland adhesive system** with one part of the gland secreting the glue that holds the animal in place and the other the releasing compound that dissolves the glue when it's time to move.

 The dual gland adhesive system is not unique to Gastrotricha; which other animal phyla have the same system?

Rotifera

Mastax and ciliated corona

The combination of a **mastax** and ciliated **corona** is unique to the rotifers; gnathostomulids have a mastax but not a corona. The corona is a large, ciliated surface composed of two parts: the anterior apical field and the circumoral or buccal field behind the apical field and surrounding the mouth. Variations on this basic plan are common, and one is enlargement of the apical field to create two **trochal discs** with cilia beating in **metachronal waves** in opposite directions on each disc. Where the water flow from the two discs meets, the cilia on the surface of the apical field propel the particulate food toward the mouth and into the pharynx. The posterior end of the pharynx forms the mastax from pharyngeal musculature and jawlike **trophi** that grind and tear at the ingested meal. The mastax is a complex structure often made up of seven trophi, and these vary in appearance depending on the rotifers food. Not all rotifers are filter feeders. In carnivorous species the ciliature on the corona is reduced and the trophi of the mastax can be used to capture prey. Crushing and grinding mastax are found in herbivores, and grasping trophi with teeth are common in carnivores. Both the corona and mastax vary between species. Their modifications are used in the taxonomy of the Rotifera.

Parthenogenesis

Of the different types of reproduction that occur in animals, **parthenogenesis**, the development of viable young from unfertilized eggs, is perhaps one of the more unusual. At first glance it might seem that such a reproductive strategy is contrary to the advantages of mixing genetic material from a male and female during mating. Many of the early animals reproduced exclusively by **asexual** mechanisms, and it was the variability that appeared in large numbers of offspring that allowed for variation in a species. In rotifers, parthenogenesis allows these minute animals to quickly colonize a favorable habitat. Mitotic divisions produce **amictic eggs** that develop into new females without fertilization. At some point females from these amictic eggs produce **mictic eggs** by meiosis, and some of the mictic eggs become males. The males produce sperm that fertilizes the remaining mictic eggs that didn't undergo parthenogenesis to produce males. The stimulus for switching from mictic and amictic egg production isn't fully understood and appears to relate to the deterioration of habitat or season. Whatever is the cause, the result is a new diploid egg that remains dormant until favorable conditions return.

✎   *What other groups of animals exhibit parthenogenetic reproduction?*

Cryptobiosis

Rotifers also exhibit an unusual survival strategy of **cryptobiosis**, an extreme state of dormancy. Unlike dormancy where there is usually some sign of life or metabolic activity, in cryptobiosis there is no sign of any life; the animals appear to be dead. Animals in this state are often able to survive extremely dry conditions, temperatures as high as 100 degrees Celsius, and temperatures as equally cold below zero. Animals in the resistant state are dehydrated, and when water becomes available they rehydrate and come back to life! Cryptobiosis is common in rotifers, tardigrads, nematodes, mites, and some insects.

Ligaments, ligament sacs, and a uterine bell (Acanthocephala)

The adult acanthocephalan is short lived, which isn't surprising because they don't have a digestive tract. A **ligament** running from behind the mouth to the cloaca at the posterior end of the animal is formed from a block of endoderm suggesting that it's a remnant of the gut. The body cavity contains either a single ligament sac or is divided in two by a pair of sacs separating the dorsal and ventral sides. Gonads are anchored to the acellular, collagenous sacs, and in the female the ovary breaks apart releasing egg masses into the pseudocoel. Fertilized eggs and embryos in the pseudocoel are at different stages of development. The **uterine bell** sorts the developing embryos and allows only those at the appropriate stage of development, at or near the **acanthus** stage, to pass to the uterus and out the female genital opening.

Spiny proboscis (Acanthocephala)

The acanthocephalans get their common name, spiny-headed worms, from their spiny **proboscis** used to anchor to the intestinal wall of their vertebrate host. When it's not extended, the proboscis rests in the **proboscis receptacle**. When the muscles surrounding the sac contract, fluid in lemnisci acts as a hydraulic medium to extend the proboscis, pulled back in by retractor muscles running from the tip of the proboscis to the base of the proboscis receptacle.

A Closer Look Inside *Digital Zoology* - Nematoda; Gastrotricha; Rotifera, including Acanthocephala

Nematoda -- Ascaris: *Pig roundworm*

If you've been reading about nematodes, one of the things you've been told is that they're supposed to be microscopic in size, and that's certainly not the case with the *Ascaris*, the pig roundworm. In part, this is because *Ascaris* is a parasitic nematode that lives in the digestive tract of its host, and presumably their size is in some way related to their parasitic life cycle. Whatever the reason, it works well for zoology students. It's much easier to see the nematode characteristics in an animal this size than one that is microscopic.

In the preserved specimens, the most obvious difference between the male and female roundworm is the hooked posterior end of the male, where the genital pore is located. You don't see that in the female, and her genital pore is midway down the body. When the two worms mate the male's body is at right angles to the female, and he uses the **spicules** at the tip of his body to hold on to the female and guide sperm into the female genital pore.

 Is there any other way that you could tell the difference between the sexes?

Before you dissect your specimen take a close look at it under the dissecting microscope, and be sure to see the anterior tip of the animal and the surface of the body wall. If you can't see the structures on the anterior end of your roundworm, take a look inside *Digital Zoology*. The mouth is surrounded by three lips and a pair of pores that lead to the amphids. This combination of lips and the position of the amphid pores is how you can tell the difference between the dorsal and ventral surfaces of the animal, with the single lip above the amphids being on the dorsal side and the other two lips below the amphids being on the ventral side of the animal.

Although at first glance you might think that the surface of the worm is smooth, with a closer look under the microscope you'll see that it's grooved and covered in ridges. Although not as important in *Ascaris* these ridges are part of how nematodes move.

The insides of a dissected specimen are deceptively simple, and you won't appreciate the complexity until you combine what you're seeing in the dissecting tray with the cross section slides. Most of the pseudocoel is filled with the reproductive tract, which looks like a tangled mass of threads. It is a tubular gonad with one end unattached and floating free in the pseudocoel, while the other end of the gonad is anchored to the genital pore. It's a long tube that winds back and forth inside the body cavity, and as it does so, its diameter gradually increases. Along its length are parts where the germ cells produce the gametes; other parts where gametes develop and mature; and finally, a place to store them before being released. Before you think you've got it sorted out, there's one last twist. The females have two of these tubular gonads and males only one.

You should be able to find where the tubes end, but the small start of the gonad may be hard to find. Other than seeing the changes in diameters, you can't tell the difference between the regions of gamete formation, maturation, and storage. That's where the cross section slides come in. In the female, the solid, wagon-wheel-like section is the ovary. The muscular oviduct without shelled eggs follows, and the paired uteri, with shelled eggs, is the largest structure you can see in the cross section. In males it's harder to tell the differences, and the size of the tube is your best guide. In most animals you would be able to identify the seminal vesicle because you would see the tails on the sperm stored there. But one of the unusual things about nematodes is that they have amoebic sperm!

Other things to match up between your dissection and the cross section are the muscles, the position of the lateral line, and the nerve cords (easier to see in the cross section). The gut is a straight tube that runs the length of the animal, and when you look at it in cross section you'll see there are no surrounding muscles.

Nematoda -- *Turbatrix: Vinegar eel*

If available, prepare a wet mount of the vinegar eel and watch how these animals move. If live specimens aren't available, video clips can be found in *Digital Zoology*. As discussed previously, the longitudinal muscles work in combination with the pseudocoel, which acts as a hydrostatic skeleton. Any change in shape of the body wall changes the shape of the whole animal. The result is the whiplike movement that you see. It doesn't look like an efficient way to get around but remember, this movement occurs in substrate and the winding movement of the body is used to move between granules of the surrounding substrate. Remember those ridges that we mentioned in the cuticle? They're important in helping the nematode push against the substrate as the nematode moves.

Gastrotricha

Gastrotrichs are one of a number of unusual phyla that inhabit the meiofauna of freshwater and marine environments. One of their distinctive features is the forked posterior end of the body with adhesive glands at the tip of each fork to hold the animal in place. A gastrotrich glides across the substrate using cilia on the ventral surface of its body, and it can bend in either direction by contracting longitudinal muscles that run the length of the body; circular muscles move the bristles and aren't involved in locomotion. The dorsal surface is covered with cuticular plates and bristles most noticeable on and around the head. The mouth is located at the anterior tip of the body and leads to the trilobed pharynx followed by an intestine. An anus opens between the two forks at the posterior end of the animal. The internal structures are often hard to see. But you may be able to locate eggs inside the reproductive system. *Digital Zoology* contains videos of gastrotrich movements including one that does a somersault letting you see the cilia on the ventral surface and the oral opening.

Rotifera -*Asplanchna and Philodina*

Rotifers are also found in the meiofauna and in both freshwater and marine environments. Asplanchna has a transparent cuticle that lets you see the internal organs. Philodina is common in cultures and prepared slides. A rotifer's body consists of three parts: an anterior head with the ciliated corona, a middle trunk, and a posterior foot. The most noticeable feature of a rotifer is the ciliated corona used for locomotion and feeding. In most prepared slides the corona is retracted, and you'll need live specimens to see how it works. If they're not available, be sure to take a look in *Digital Zoology* to see the metachronal waves created by cilia on trochal discs and the movements of the mastax that grind the ingested food. Asplanchna is an unusual rotifer with an incomplete gut, and it broods its developing young; you can see a miniature adult through the cuticular wall.

Rotifera - Acanthocephala

*D*igital Zoology uses prepared microscope slides to look at these specialized parasitic rotifers. The spiny proboscis is easy to see, as are the **lemnisci** that contain the hydraulic fluid that extends the proboscis; retractor muscles pull it back in. Adult acanthocephalans are parasites of the vertebrate digestive tract, and the ligaments visible inside the specimen are all that remains of the digestive system. Nutrients are absorbed across the body wall covered in a thin cuticle and syncytial epidermis that has large and easy-to-see nuclei. The reproductive system is the most prominent internal feature in both males and females. Males are usually smaller than females, and the paired testes are stacked one on top of the other. Posterior to these are the **cement glands** and the cement reservoir. In the female either a single or paired ovary develops on the surface of the ligament and early in the female's life breaks up forming egg masses that float in the pseudocoel. Sperm from the male passes into the female's pseudocoel, and eggs and embryos at different stages are found floating inside the female pseudocoel. The ligament connects to a uterine bell that sorts the embryos and only lets those at the appropriate stage in development past the bell into the uterus and into the feces of the host.

Cross-References to Other McGraw-Hill Zoology Titles

Integrated Principles of Zoology, 11th edition. C. P. Hickman, L. S. Roberts, and A. Larson. Chapter 15.

Animal Diversity, 3rd edition. C. P. Hickman, L. S. Roberts, and A. Larson. Chapter 9.

Zoology, 5th edition. S. A. Miller and J. P. Harley. Chapters 10 and 11.

Biology of the Invertebrates, 4th edition. J. Pechenik. Chapters 11, 16 and 18.

Laboratory Studies in Integrated Principles of Zoology, 10th edition. C. P. Hickman, F. Hickman, and L. Kats. Chapter 10.

General Zoology Laboratory Guide, 13th edition. C. Lytle. Chapter 10.

Structures Checklist

Here are some of the structures that you should be able to easily find in *Digital Zoology* and the specimens that you will be looking at in your lab. After reading your lab handout, you might want to add more and, depending on the equipment available in your lab, you might see more. As you study the material, you might also want to make some notes on how some of these structures looked or include a drawing in your lab notes. (Structures indicated by * may be hard to see.)

Ascaris

Preserved specimen - intact and dissected

- ☐ Amphid pores
- ☐ Copulatory spicules
- ☐ Intestine
- ☐ Lateral line
- ☐ Lips
- ☐ Mouth
- ☐ Ovary
- ☐ Oviduct
- ☐ Pharynx

- ☐ Seminal vesicle
- ☐ Sperm duct
- ☐ Testis
- ☐ Uterus
- ☐ Vagina
- ☐
- ☐
- ☐

Cross section

- ☐ Cuticle
- ☐ Dorsal nerve cord

- ☐ Epidermis
- ☐ Intestine
- ☐ Lateral line (Renette cells)
- ☐ Longitudinal muscle
- ☐ Ovary
- ☐ Oviduct
- ☐ Seminal vesicle
- ☐ Sperm duct
- ☐ Testis
- ☐ Uterus

- ☐ Ventral nerve cord
- ☐
- ☐
- ☐

Additional structures

- ☐
- ☐
- ☐

Gastrotricha

- ☐ Anus
- ☐ Brain*
- ☐ Forked tail

- ☐ Intestine
- ☐ Mouth
- ☐ Ovary

- ☐ Pharynx
- ☐ Sensory bristles
- ☐

- ☐

Rotifera, including acanthocephala

Rotifer

- ☐ Antenna
- ☐ Cilia
- ☐ Cingulum
- ☐ Cloaca*
- ☐ Corona
- ☐ Eggs*
- ☐ Embryos
- ☐ Foot
- ☐ Gastric gland
- ☐ Germovitellarium
- ☐ Intestine
- ☐ Mastax

- ☐ Nephridial duct
- ☐ Pedal cement glands
- ☐ Stomach
- ☐ Toes
- ☐ Trochal disc
- ☐ Vitelline gland
- ☐
- ☐
- ☐
- ☐
- ☐
- ☐
- ☐
- ☐

Acanthocephala

- ☐ Anterior testis
- ☐ Bell pocket
- ☐ Cement gland
- ☐ Female gonopore
- ☐ Giant nuclei
- ☐ Lemnisci
- ☐ Ligament
- ☐ Ligament sac
- ☐ Male gonopore
- ☐ Ovary
- ☐ Posterior testis
- ☐ Proboscis

- ☐ Pseudocoel
- ☐ Saeftigen's pouch
- ☐ Seminal vesicle
- ☐ Uterine bell
- ☐ Uterus
- ☐ Vagina
- ☐
- ☐
- ☐
- ☐

Crossword Puzzle – Nematoda; Gastrotricha; Rotifera, including Ancanthocephala

Puzzle solution. An interactive web-based version of this puzzle and its solution are available on the *Digital Zoology* website www.digitalzoology.com/students. With the interactive puzzle you can check to see if individual words or the whole puzzle is correct and get hints for single letters.

Across

1 Larval acanthocephala are parasites of animals in this phylum. (10)

6 These sensory structures surround a nematode's mouth. (7)

12 The number of muscle layers in a nematode. (3)

13 These glands help a rotifer stick to the substrate. (5)

16 These osmoregulatory structures are more important in freshwater species than marine rotifers. (14)

18 A nematode has this number of ovaries. (3)

19 The fluid in the pseudocoel is a part of this type of skeleton. (11)

20 Long thought a separate phylum, we now know they are parasitic animals from this phylum. (8)

22 These create the water currents a rotifer uses for feeding. (5)

23 When a rotifer's outer cuticle is solid and covering the trunk, it's called this. (6)

26 Anterior end on nematodes appears to have this type of radial symmetry. (10)

28 Unlike most males that have two, male nematodes have only one of these. (6)

31 Describes the gut of an acanthocephalan. (7)

32 Animals in the phylum Nematoda are commonly called this type of worm. (5)

33 Where fertilization occurs in the body of an acanthocephalan. (10)

34 The internal organs of a nematode float in the body cavity because it does not have any of these to anchor them in place. (11)

36 The outer cuticle of a nematode is composed of this protein. (8)

38 Unlike other animals, the cilia of a gastrotrich are covered in this. (7)

39 Extensions of these cells connect the nervous and muscular systems of a nematode. (6)

40 Nematodes have this type of a digestive system. (8)

Down

1 The type of movement used by nematode sperm. (7)

2 The number of nematode lips. (5)

3 Free-living nematodes are an important part of this nutrient cycle. (13)

4 Dormant rotifer eggs can withstand conditions of high heat, extreme cold, and/or desiccation; a process referred to as this. (12)

5 In nematodes the opening to the female genital system is found in this region of the body. (6)

7 *Ascaris* is an unusually large nematode because of its adaptation to this type of life. (9)

8 The coelomic state of a gastrotrich. (10)

9 If Rotifera and Acanthocephala become one phylum, it will probably be called this. (10)

10 Gastrotrichs are so small we have a special name for them and other animals like them. (9)

11 Describes the epidermis of an acanthocephalan. (9)

14 These help acanthocephalans hold onto their vertebrate host. (6)

15 Most rotifers of a species have the same number of cells; it's an example of this. (6)

17 Freshwater gastrotrich populations usually are missing males; this describes their reproductive strategy. (15)

21 The first larval stage in the acanthocephalan life cycle. (8)

24 This lets only embryos at the proper stage of maturation enter the uterus of a female acanthocephalan. (7,4)

25 You'll find an adult acanthocephalan living inside this part of a vertebrate. (9)

27 Freshwater rotifers often use the reproductive strategy of parthenogenesis; as a result, these are usually missing in the population. (5)

29 The spinning wheels on a rotifer's head. (6)

30 The side of a gastrotrich where you'll find the cilia. (7)

31 The curl on the posterior end of a nematode identifies its gender as this. (4)

35 Nematodes are never found in the plankton because they don't do this very well. (4)

37 In a nematode the endoderm forms this structure. (3)

Self-Test – Nematoda; Gastrotricha; Rotifera, including Acanthocephala

Use the following labels to identify the photographs. You may have to use a label more than once, and some labels may not be appropriate for the photographs. Answers are available on the *Digital Zoology* website at www.digitalzoology.com/students. Be sure to try the interactive drag-and-drop quizzes on the *Digital Zoology* CD-ROM.

Specimen

A-Nematode
B- Acanthocephalan
C- Rotifer

Labels

1) Antenna
2) Bursa
3) Cement gland
4) Corona
5) Embryo
6) Gastric glands
7) Gut lumen
8) Gut wall
9) Lateral line
10) Lemnisci
11) Ligament
12) Longitudinal muscle
13) Mastax and musculature
14) Ovary
15) Oviduct
16) Proboscis
17) Pseudocoel
18) Stomach
19) Testis
20) Uterus
21) Vitelline gland

ARTHROPODA

Inside *Digital Zoology*

As you explore the Arthropoda on the *Digital Zoology* CD-ROM or on the web at www.digitalzoology.com, don't miss these learning tools:

- ○ Photos of intact and dissected specimens of the crustacean crab (*Cancer*) and crayfish (*Cambarus*) and the uniramian grasshopper (*Romalea*) and cockroach (*Periplaneta*).
- ○ Video clips of the circulatory system of the cockroach showing the heart beating and blood flowing through the veins of the wing.
- ○ Drag-and-drop quizzes on the key internal and external structures of the crab, crayfish, cockroach and the external anatomy of the grasshopper.
- ○ Interactive cladograms for the three arthropod subphyla: Chelicerata, Uniramia, and Crustacea. A summary of the key characteristics of each class are combined with an interactive glossary of terms.

Defining Differences

Some of the differences described in the following sections are features appear for the first time in the Arthropoda, and they define the phylum. Others are important for how animals in the phylum function. Arthropods are the most abundant animal phylum on earth, and they make that claim because of the tremendous numbers of insects included in the subphylum Uniramia. There's a lot to look at, and the features we've listed apply to all arthropods, although in many of the subphyla they have been extensively altered and may even have disappeared. Whichever is the case, you'll want to watch for examples of these in the various arthropod specimens that you will be examining.

- ○ Tagmatization of the
- ○ chitinous exoskeleton, growth by
- ○ molting,
- ○ muscle arranged in bands,
- ○ jointed appendages, and
- ○ compound eyes.

Tagmatization

An annelid's body is made up of a series of repeating units sequentially arranged down the length of the animal. **Metamerisation** describes the process that resulted in the segments, **metameres**, each of which is identical to the one on either side of it. Arthropods took metamerisation a step farther. **Tagmatization** grouped adjacent metameres into larger functional units, **tagma**, responsible for performing specialized tasks. In addition to how the segments were organized into tagma, another important difference between the two phyla is that arthropods have paired appendages on each their metameres.

A word of warning! This sequence of events assumes that arthropods were the logical evolutionary step that followed annelids. There is growing evidence that this may not be the case. This was discussed earlier in the workbook when we looked at the first animals with a cuticle, the nematodes. If the **Ecdysozoa** is a valid **monophyletic** taxon, then this annelid-to-arthropod transition is inappropriate. While we wait for a resolution to this taxonomic conundrum, the transition from a metameric body plan to one based on tagma is probably correct. The question is whether an annelid is the right metameric ancestor.

To understand the impact tagmatization has on how an animal functions, let's take a look at the ancestral and modified body plans of two different arthropod subphyla: Crustacea and Uniramia.

Primitive crustaceans had only two tagma, a **head** and a **trunk**. The head was formed from five segments and their five pairs of appendages. This included two pairs of sensory appendages: **antennae** and **antennules,** and three pairs of feeding appendages: **mandibles, first maxillae,** and **second maxillae** used for grinding, cutting, and manipulating food before it was ingested. Each of the trunk appendages had three responsibilities: locomotion, respiration, and food gathering. As this primitive crustacean paddled along using its trunk appendages, it trapped particulate food using **setal hairs** on the surface and passed it forward to the mouth. While all this was happening, the large surface area of the legs was used for gas exchange. Compare that to the lobster or crayfish. The head tagma has been replaced with a **cephalothorax.** Some of the trunk appendages have been recruited to help with feeding and **maxillipeds** are part of the "cephalo" part of the cephalothorax. The "thorax" part, more correctly the **pereon** with its **pereopods,** is locomotary and houses the main organ systems. The rest of the trunk, **pleon** with **pleopods,** is a muscular tail used to escape from predators and in the females for brooding their eggs.

In the Uniramia centipedes, the head tagma is a fusion of six segments followed by the trunk tagma with a pair of walking legs on each of the segments. The uniramian head hasn't changed much between centipedes and insects. It processes sensory information from the antennae and **compound eyes** and collects and manipulates food before it's ingested. The biggest difference can be seen in changes to the trunk, which in insects is now the **thorax** and **abdomen.** This second insect tagma, the thorax, is specialized for locomotion, and the only thing you'll find inside it are muscles that move the wings and legs. The organ systems pass their ducts and tubes through the thorax on their way to the third tagma, the abdomen, which houses the main organ systems. The abdomen's role is the day-to-day functioning of the animal, and it contains the reproductive system.

 We've described the ancestral tagma of two of the arthropod subphyla; how are the tagma and appendages arranged in a Chelicerata?

As the tagma take on these different roles, their shapes, and the shapes of the appendages attached to them, change. Arthropod cuticle is a "living plastic" and can be molded into almost anything you could think of. The result is a tremendous array of different cuticular shapes and forms, many of which are arthropod tools. For example, cuticular appendages can be delicate sensory antennae, shovels, scissors, grinding stones, walking legs, defensive pincers, and paddles for swimming. The possibilities are almost endless, and it's one of the reasons why arthropods are such a successful group.

Chitinous exoskeleton

Invertebrate cuticles are made from either **collagen** or **chitin.** Arthropods use chitin, a long chain carbohydrate polymer of modified glucose units, **N-acetyl-glucosomaine** to be more precise. The use as a structural fiber is not new with the arthropods. The cell walls in plants are strengthened with cellulose fibers, chains of glucose units. Like arthropods, fungi use chitin in their cells.

Arthropod cuticle has two important functions. The first is a combination of strength and support, and the **exoskeleton** creates a suit of armor that protects the animal inside. The cuticle's other function, as a chemical barrier, may not be as obvious. The exoskeleton prevents things from diffusing out of the animal or unwanted substances from diffusing in. This is a particularly important role for terrestrial arthropods, which need to keep water inside.

The strength of the arthropod's cuticle is a mix of chitin and cuticular proteins. This mix is solidified in the **procuticle** by a combination of chemical cross-linking and/or embedded calcium salts. Chelicerates and uniramians use chemical cross-linking and crustaceans use calcium salts; it's why when you crack into your lobster claws or crab legs you hear that crusty, crunching sound.

The chemical barrier is created by the outer **epicuticle,** and in terrestrial arthropods it involves a complex mixture of proteins and waxes that waterproof the animal. Arthropods were the first animals to conquer the terrestrial environment, and they have the most complex epicuticles. Waterproofing the cuticle was an important event in the evolution of terrestrial arthropods.

 Why don't crustaceans have a waterproof epicuticle? What's the consequence of not having one?

By varying the depth of procuticle, and/or the degree of cross-linking or embedded salts, the cuticle can be thick and hard or thin and membranous. Hard plates, **sclerites**, are connected to each other by membranous areas, which allows for bending and movement of the arthropod's armor.

Molting

There's one problem with living in a suit of armor. As the animal grows it needs to discard its current shell for a new, larger suit of armor. That's what arthropods do, and the process is referred to as **ecdysis** or **molting**. Building a suit of armor was a major metabolic investment for an arthropod. There's more to molting than discarding the old cuticle. Arthropods recycle parts of their old cuticle to help build the new one. **Chitinases** breakdown the chitin and **proteinases** attack the proteins to regenerate N-acetyl-glucosamine and the amino acid subunits that help to build the new cuticle.

 Why do the chitinases and proteinases breakdown only the old cuticle and not the new one?

Muscle arranged in bands

A solid outer cuticle has affected the way that muscles are arranged in arthropods. Until this point in our discussions of different animals, the **hydrostatic skeletons** have been the most common type of skeleton. Sheets of muscles are arranged in opposite directions to each other, circular and longitudinal, and surround a fluid-filled space. Even nematodes, with their cuticle, still had a way to make it flexible enough to be used as hydrostatic skeleton for movement. The arthropod's cuticle isn't like a nematode's; it's solid, making the use of a hydrostatic skeleton impossible. Sheets of muscles are replaced by bands of muscle embedded in solid cuticle on either side of a membranous piece of cuticle. As the muscle contracts and shortens, it moves one piece of cuticle relative to the other. This type of muscular arrangement, in combination with the tubular exoskeleton, creates the unique jointed limbs and feet that give the phylum its name, the Arthropoda – jointed foot animals.

Jointed appendages

In its simplest form, an arthropod appendage is a tubular shell of cuticle surrounding muscles. There is only one type of joint that can be used to make this bend, and the two tubes on either side of the joint come in contact with each other at two points, the hinges or pivot points. When one tube moves relative to the other it bends in only one direction, pivoting around the line between the two hinges. Having only one of these joints on an appendage isn't enough. Farther along are more of these joints, each of which bends at a different angle. With each joint bending in its own direction, the tip of the appendage can be positioned anywhere within its reach.

We started with the analogy that an insect's exoskeleton was like a suit of armor and joints in the arms and legs of a suit of armor work the same way that they do in arthropods. From ancient armor, to robotic arms on an automobile assembly line, to the arm on the space shuttle or space station, they all have the same design constraints as arthropod appendages.

There are biomechanical limits to how big structures can become with this type of tubular construction. As the appendage becomes larger, the wall of the tube has to become thicker to withstand forces generated by the larger muscles inside. This increasing thickness means increased weight. There's a point where the muscle inside the tube can't lift the weight of the appendage. The way around the problem is to place the muscles on the outside of a solid supporting skeletal rod, an endoskeleton—a strategy used by the chordates.

 Marine crustaceans and terrestrial arthropods have the same tubular skeleton and appendages. Why are the crustacean's appendages so much larger?

Compound eyes

Arthropods have a unique visual system, the compound eyes. Or at least most of them do; it is missing in many of the chelicerates. Presumably, its absence is an advanced trait because compound eyes are found in primitive horseshoe crabs, also chelicerates.

The compound eye is made up of a series of optic units, the **ommatidia,** the fundamental units for creating the image. Each ommatidia is formed from a series of light-sensitive **retinular cells** optically isolated from each other by **pigment cells**. Different retinular cells are responsive to different wavelengths of light, and their responsiveness corresponds to three primary colors. The cells detect different intensities of color as they interpret the small part of the bigger picture that it samples. Each of these individual pieces of the picture is combined by the arthropod brain to create the final image.

Think of each of the ommatidia as a pixel on a computer screen or television. The retinular cells input a value for each of the red, green, or blue color depths of the pixel. You can see images and motion on the screen as the pixels change their colors. The arthropod eye works much like this, with one notable exception with insects. Insects don't see the color red at the upper end of the color spectrum, but do see ultraviolet at the lower end of the spectrum.

Although the compound eye is usually identified as an arthropod trait, arthropods also have **simple eyes** or **ocelli**. The simple eyes don't usually form images and are usually involved in detecting light intensity. A notable exception to this are the large modified simple eyes of hunting spiders. The loss of the image forming compound eye early in the evolution of the chelicerates left these spiders with only the simple eyes that could be modified for forming images.

✍ *Eyes are important in detecting and capturing prey. How do chelicerates do that without compound eyes?*

A Closer Look Inside *Digital Zoology* - Arthropoda

Crustacea: Crayfish and crab

One of the traditional lab exercises in zoology is to examine **serial homology** in the legs of a crayfish or lobster. (*Digital Zoology* also has a crab that you can do this with as well.) One of the differences between annelid and arthropod **metameres** is arthropod segments have a pair of appendages. In chelicerates and uniramians, they are **uniramous**; crustacean appendages are **biramous**. Each appendage has two branches, an outer **exopodite** and an inner **endopodite** attached to a single **protopodite** that anchors the limb on the body. The protopodite may have cuticular extensions on their inside surface, **endites**, or the outer surface, **exites**. Any, or all of these, can be modified in different ways for locomotion, feeding, or respiration. As you examine each of the appendages in your lab, or with *Digital Zoology*, you'll see an example of the "living plastic" of arthropod cuticle and how it can be molded into a variety of shapes. That's the reason in almost every zoology course you'll find students pulling the appendages of their specimen and lining them up on the bench – it's almost a right of passage.

Crustaceans have an **open circulatory system,** and their **heart** is located inside a **pericardial cavity** on the dorsal side of the animal. Crab and crayfish blood contains the respiratory pigment **hemocyanin,** and it's essential that all tissues receive oxygenated blood. Three major blood vessels supply blood to the anterior (**anterior aorta**), posterior (**posterior aorta**), and ventral surface (**sternal artery**) of the animal. Each vessel divides further into smaller vessels that supply blood to distinct organs or parts of the body. There are no capillaries in the system, and oxygenated blood is poured over the organs and tissues and then pools in the **hemocoel**. Here it continues to supply oxygen and nutrients, and it picks up metabolic wastes from the structures that it bathes. As the crustacean heart beats it pulls blood from the hemocoel through the **gills** and out through the three major blood vessels. Gills in these large crustaceans are housed inside **branchial chambers. Gill bailers** help to keep water moving across the respiratory surfaces.

✍ *How do crabs and crayfish eliminate metabolic wastes?*

Crayfish and crabs have an unusual **gastric mill**, or **stomach,** as part of their digestive system. What makes it unusual are the three hard chitinous teeth inside. Students often wonder how there can be teeth so far down the digestive tract. The answer is in how the gut forms in arthropods and other animals. The gut forms from **ectodermal** invaginations from the anterior and posterior end of the embryo that join with the **endoderm** in the middle. This creates the **archenteron**, or primitive gut. An arthropod's ectoderm is lined with cuticle, and the ectodermal front and back parts of the gut have a cuticular lining. The gastric mill is part of the ectodermal lining. There is, however, a potential problem with having a cuticle lined gut, and it's related to the cuticle's role as a chemical barrier. Digestive enzymes can't be secreted across the cuticle, and nutrients can't be absorbed, either. That has to be done by the part of the gut formed from the endoderm. In these two animals, this is the job of the endodermal **digestive gland** and the **midgut,** which has no cuticular lining.

✍ *How does locomotion differ in the crab compared to crayfish or lobsters?*

You shouldn't be left with the impression that all crustaceans are the size of the two specimens in *Digital Zoology*. The vast majority of crustaceans are microscopic zooplankton of aquatic environments. These minute crustaceans occupy a critical position in marine food chains as the primary herbivores in that environment.

Uniramia: Cockroach and grasshopper

Over 85 percent of the animals on the planet, over a million species, are insects. They all look remarkably similar to each other. If it's an arthropod with three pairs of legs, it's an insect. You might be tempted to include wings in the definition but primitive insects were wingless making wings a **derived trait**, rather than an **ancestral trait** for the group. There are a number of reasons why insects are so successful. One of the more important was their being the first animals to successfully colonize land. They were able to do this because they found a way to waterproof their bodies and conserve water. One way they did this was through a structural modification to the cuticle; the other was a functional change that resulted in a unique excretory system – **Malpighian tubules**. More precisely, it should be referred to as the **Malpighian tubule hindgut complex** because **rectal pads** in the **hindgut** are responsible for absorbing the last of the water from the fecal material before it leaves the animal.

Insects have an extra **waxy layer** in their **epicuticle** that waterproofs the whole surface of the animal. It's combined with a complex mixture of cross-linked proteins, the **cement layer**, that hardens the surface against abrasion, another potential problem in the terrestrial environment.

Uric acid is the main metabolic waste in insects. It is not very toxic, and it packs more nitrogen per molecule than either **urea** or **ammonia**. Malpighian tubules are closed at one end and empty into the gut at the junction between the **midgut** and **hindgut**. The tubules are bathed in hemolymph and create an isotonic **primary urine**, a hemolymph ultrafiltrate of the hemolymph containing molecules small enough to pass through the membrane of the tubule cells: water, salts, low-molecular weight nutrients, and dissolved uric acid. This is mixed with the undigested food as it leaves the midgut, and the hindgut absorbs water, salts, and nutrients contained in the mix of fecal material and urine and leaves the uric acid in the rectum. Depending on where an insect lives, and how limited water is, the hindgut absorbs the appropriate amount of water. Insects that live inside seeds for their whole life never have the opportunity to drink water, and their Malpighian tubule hindgut complex is so efficient at recovering water that metabolic water generated by metabolic process is sufficient for the insect to survive. The fecal material is a dry dust.

Insects have an open circulatory system, and their dorsal heart pumps blood toward the head. From there it percolates back through the large hemocoel bathing the organs and structures housed inside the cavity. Unlike crustacean blood, or the blood from most of the other animals, insect blood is not involved in oxygen transport. Insects have a **tracheal system** that supplies air directly to the tissues. **Spiracles** open in the side of the thorax and abdomen and are connected to **tracheal trunks** strengthened by rings of cuticle, **tenidia**, which prevent the trunks from collapsing. Their tracheal trunks connect to **trachea**, which branch throughout the insect's body. The large surface area created by the tracheal system is a potential site for water loss. That doesn't happen, however, because the trachea are covered with the waxy epicuticle. Gas exchange occurs across the smaller **trachioles**, which don't have this waterproof covering. The tracheole's microscopic fingers extend and wrap around the cells into which oxygen diffuses. With the blood no longer involved in oxygen transport, it functions more as a tissue culture medium supplying tissues with nutrients and materials they need and carrying away the metabolic wastes. Insect blood is high in protein, sugars, fats, and amino acids.

Cross-References to Other McGraw-Hill Zoology Titles

Integrated Principles of Zoology, 11th edition. C. P. Hickman, L. S. Roberts, and A. Larson. Chapters 18-20.

Animal Diversity, 3rd edition. C. P. Hickman, L. S. Roberts, and A. Larson. Chapter 12.

Zoology, 5th edition. S. A. Miller and J. P. Harley. Chapter 14 & 15.

Biology of the Invertebrates, 4th edition. J. Pechenik. Chapter 14.

Laboratory Studies in Integrated Principles of Zoology, 10th edition. C. P. Hickman, F. Hickman, and L. Kats. Chapter 13-15.

General Zoology Laboratory Guide, 13th edition. C. Lytle. Chapter 13.

Structures Checklist

Here are some of the structures that you should be able to easily find in *Digital Zoology* and the specimens that you will be looking at in your lab. After reading your lab handout, you might want to add more and, depending on the equipment available in your lab, you might see more. As you study the material, you might also want to make some notes on how some of these structures looked or include a drawing in your lab notes. (Structures indicated by * may be hard to see.)

Crayfish

External anatomy
- ☐ Abdomen
- ☐ Antennae
- ☐ Antennules
- ☐ Anus
- ☐ Branchial chamber
- ☐ Carapace
- ☐ Cepahlothorax
- ☐ Cervical groove
- ☐ Cheliped
- ☐ Compund eye
- ☐ First maxilla
- ☐ First maxilliped
- ☐ Gonopods (male swimmeret)
- ☐ Mandible
- ☐ Mouth
- ☐ Opening of antennal gland
- ☐ Rostrum
- ☐ Second maxilla
- ☐ Second maxilliped
- ☐ Sternum

- ☐ Swimmerets (pleopods)
- ☐ Telson
- ☐ Uropods
- ☐ Walking legs
- ☐
- ☐
- ☐
- ☐
- ☐

Dissection
- ☐ Abdominal extensor muscle
- ☐ Abdominal flexor muscle
- ☐ Antennal artery
- ☐ Antennal gland
- ☐ Anterior gastric muscle
- ☐ Digestive gland
- ☐ Dorsal abdominal artery

- ☐ Esophagus
- ☐ Gastric mill, pyloric region
- ☐ Gastric mill, cardiac region
- ☐ Gastric teeth
- ☐ Gill bailer (scaphognathite)
- ☐ Gills, arthrobranchs
- ☐ Gills, podobranchs
- ☐ Heart
- ☐ Heart ostia
- ☐ Hepatic artery
- ☐ Intestine
- ☐ Mandibular muscle
- ☐ Nerve cord
- ☐ Ophthalmic artery
- ☐ Ovary
- ☐ Oviduct*
- ☐ Pericardial cavity
- ☐ Posterior gastric muscle
- ☐ Sperm ducts*

- ☐ Sternal artery
- ☐ Supraesophagial ganglion (brain)*
- ☐ Testes
- ☐ Ventral abdominal artery
- ☐ Ventral thoracic artery
- ☐
- ☐
- ☐
- ☐

Additional structures
- ☐
- ☐
- ☐
- ☐
- ☐
- ☐

Crab

External anatomy
- ☐ Abdomen
- ☐ Antennae
- ☐ Antennules
- ☐ Cephalothorax
- ☐ Cheliped
- ☐ Compound eye
- ☐ Exhalan opening to gill chamber
- ☐ First maxilla
- ☐ First maxilliped
- ☐ Flabella on maxillipeds
- ☐ Gill bailer (scaphognathite)

- ☐ Inhalant opening to gill chamber
- ☐ Mandible
- ☐ Mouth
- ☐ Pleopods
- ☐ Second maxilla
- ☐ Second maxilliped
- ☐ Sub-branchial carapace
- ☐ Third maxilliped
- ☐ Thoracic sterna
- ☐ Walking legs
- ☐
- ☐
- ☐
- ☐

Dissection
- ☐ Antennal glands*
- ☐ Anterior gastric muscle
- ☐ Digestive gland
- ☐ Gastric mill, cardiac region
- ☐ Gastric mill, pyloric region
- ☐ Gill cleaner
- ☐ Gills
- ☐ Heart
- ☐ Heart ostia
- ☐ Mandibular muscle
- ☐ Ovary

- ☐ Oviduct
- ☐ Pericardial cavity
- ☐ Posterior gastric muscle
- ☐ Sperm duct
- ☐ Testes
- ☐
- ☐
- ☐

Additional structures
- ☐ Dimorphism of the male and female abdomen
- ☐
- ☐

Grasshopper and Cockroach

External anatomy

- ☐ Abdomen
- ☐ Antennae
- ☐ Arolium
- ☐ Cerci (cockroach only)
- ☐ Claw
- ☐ Compound eyes
- ☐ Coxa
- ☐ Femur
- ☐ Forewing
- ☐ Head
- ☐ Hindwing
- ☐ Hypopharynx
- ☐ Labial palps
- ☐ Labium
- ☐ Labrum
- ☐ Legs
- ☐ Mandible
- ☐ Maxilla
- ☐ Maxillary palps
- ☐ Mesothorax
- ☐ Metathorax
- ☐ Ovipositor
- ☐ Prothorax
- ☐ Simple eyes, ocelli
- ☐ Spiracles
- ☐ Sternites
- ☐ Tarsomeres
- ☐ Tarsus
- ☐ Tergites
- ☐ Tibia
- ☐ Trochanter
- ☐ Tympanum (grasshopper only)
- ☐
- ☐
- ☐
- ☐
- ☐
- ☐

Dissection

- ☐ Abdominal ganglia
- ☐ Digestive ceaca
- ☐ Esophagus
- ☐ Fat body
- ☐ Female accessory glands
- ☐ Flight muscles
- ☐ Heart
- ☐ Male accesory glands
- ☐ Malpighian tubules
- ☐ Midgut
- ☐ Nerve cord
- ☐ Ovary
- ☐ Oviduct
- ☐ Rectum
- ☐ Salivary glands
- ☐ Sperm duct*
- ☐ Stomach
- ☐ Subesophageal ganglion*
- ☐ Supraesophageal ganglion (brain)*
- ☐ Testes
- ☐ Thoracic ganglia
- ☐ Trachea
- ☐ Tracheal trunks
- ☐
- ☐
- ☐
- ☐
- ☐

Additional structures

- ☐ Immature stages of the life cycle
- ☐ Ootheca of the cockroach
- ☐
- ☐
- ☐
- ☐

Crossword Puzzle – Arthropoda

Puzzle solution. An interactive web-based version of this puzzle and its solution are available on the *Digital Zoology* website www.digitalzoology.com/students. With the interactive puzzle you can check to see if individual words or the whole puzzle is correct and get hints for single letters.

Across

2 Spiders use this to build their webs. (4)

4 Although chelicerates don't have compound eyes, you'll still find this type of light-sensing structure on most species. (6,3)

7 The Malpighian tubule can't complete the excretory process without this other part of the gut. (6)

11 Arthropods have this type of circulatory system. (4)

13 The most anterior appendage in a chelicerate. (8)

14 A male spider taps the female because he wants to do this. (4)

16 Coevolution of insects with this group of organisms created much of the biodiversity that we now see on the planet. (6)

17 Another name for the crayfish stomach. (7,4)

21 In Crustacea, the compound eyes are found on the tip of these structures. (4,5)

22 Smaller than trachea, this part of the tracheal system permits gas exchange with the surrounding tissues. (10)

25 Based on neurological evidence, the uniramian head consists of this many segments. (3)

26 Air enters an insect's tracheal system through these. (9)

27 When first secreted, the molting fluid is in this form and unable to digest the old cuticle. (8)

28 Is it true that even with an open circulatory system, crayfish have veins and arteries? (3)

29 A terrestrial insect may swallow this to help escape from the old cuticle. (3)

32 Sternites are found on the ventral surface of this insect tagma. (7)

34 In addition to large eyes, insects also have these eyes. (6)

38 The feeding appendage found in Uniramia and Crustacea. (8)

40 The Arthropoda is considered to be a very successful group in part because of its _____ evolutionary history. (4)

41 Spiders inject their food with digestive enzymes, wait, and then drink up the liquefied food. It's a special type of digestion. (14)

42 The part of the light spectrum an insect can see that human beings can't. (2)

44 Insects don't see this color in the spectrum. (3)

47 Both protein and this substance are digested and reused during molting. (6)

48 The name of the openings in the crustacean hearts. (5)

49 Male insects pass these sperm packages to the female insects. (14)

50 Many chelicerates immobilize their prey by injecting this. (5)

51 Number of appendages on the first tagma of a chelicerate. (5)

52 Extra corporeal digestion means that for chelicerates, most of the digestion of captured prey occurs where? (7)

Down

1 Because their surface looks like pages of paper, the respiratory structures of spiders are referred to as these. (4,5)

2 Insects protect the egg from drying out by covering it with this. (5)

3 This group of small crustaceans are found in the millions and are fed on by larger animals in the marine environment. (5)

5 The miniature ectoparasitic chelicerates. (5)

6 A more technical term for molting, which also describes the escape of an arthropod from its old cuticle. (7)

8 Insects in which the adults and their young look completely different from each other, are examples of this type of metamorphosis. (8)

9 A male insect's sperm package. (13)

10 The basic sugar unit of the complex carbohydrate found in insect procuticle. (11)

12 Wings are found only in this stage of the insect life cycle. (5)

15 Blood in the dorsal vessel of an insect is pumped toward this part of the animal. (8)

17 Large crustaceans use these to breathe; it's not a problem for the smaller ones who use simple diffusion. (5)

18 The coxal glands of spiders are replaced with these tubules in some terrestrial species. (10)

19 These compounds in the epicuticle give it its waterproof character. (5)

20 The model describing arthropod vision. (6,6)

23 The cephalothorax results from the fusion of some of the thoracic segments with this tagma. (4)

24 This is the lower of the two cuticular plates that surrounds the crayfish abdomen. (7)

30 When a thoracic leg in a crustacean becomes involved in feeding, it is referred to as this type of appendage. (10)

31 The unique image-forming sensory apparatus of an arthropod is this type of an eye. (8)

33 Can molting fluid digest the exocuticle? (2)

34 This is inside a statocyst. (9)

35 The antennal gland is involved in this function. (9)

36 Spider webs are used to trap these. (7)

37 Uniramians have this number of sensory appendages on their head. (3)

39 An insect's upper lip. (6)

42 The most posterior appendage on a crayfish. (6)

43 Chelicerates have only this many tagma. (3)

45 How many teeth are there in the stomach of a crayfish? (5)

46 Because the grinding surface of the crustacean mandible is located here on the mandible, we refer to it as a gnathobasic mandible. (4)

Self-Test - Arthropoda

Use the following labels to identify the photographs. You may have to use a label more than once, and some labels may not be appropriate for the photographs. Answers are available on the *Digital Zoology* website at www.digitalzoology.com/students. Be sure to try the interactive drag-and-drop quizzes on the *Digital Zoology* CD-ROM.

Specimens

A- Crayfish

B- Cockroach

C- Grasshopper

Labels

1) Abdomen
2) Abdominal artery
3) Abdominal muscle
4) Antenna
5) Branchial chamber
6) Cephalothorax
7) Cercus
8) Cheliped
9) Collaterial glands
10) Compound eye
11) Crop
12) Digestive ceca
13) Digestive gland
14) Fat body
15) Femur
16) Heart
17) Hindgut
18) Intestine
19) Malpighian tubules
20) Mesothoracic leg
21) Midgut
22) Pronotum
23) Prothoracic leg
24) Tarsus
25) Tibia
26) Uropod
27) Walking leg
28) Wing

BRYOZOA AND BRACHIOPODA

Inside *Digital Zoology*

As you explore the Bryozoa and Brachiopoda on the *Digital Zoology* CD-ROM or on the web at www.digitalzoology.com, don't miss these learning tools:

❍ Photos of prepared slides of the marine (*Bugula*) and freshwater (*Pectinatella* and *Plumatella*) bryozoans and the brachiopod *Lingula*.
❍ Drag-and-drop quizzes on the key structures of bryozoans and brachiopods.
❍ Interactive cladogram showing the major events that gave rise to the main deuterostome groups. A summary of the key characteristics of each of the deuterostome phyla and subphyla of chordates and hemichordates are combined with an interactive glossary of terms.

Defining Differences

Bryozoans, brachiopods, and phoronids are referred to as the lophophorate animals because they have a lophophore. Some of the features in the following list are common to all three; others are unique to the individual phyla. Other characters in the list are things that we see for the first time in animals and are important for understanding how the animals in these phyla function. Whichever is the case, you'll want to watch for examples of these in the specimens you'll be examining. In *Digital Zoology* we look at bryozoans and brachiopods as examples of lophophorates.

❍ Tripartite coelom,
❍ lophophore, and
❍ U-shaped gut.

Tripartite coelom

Deuterostomes share a number of characteristics: **radial cleavage** of the fertilized egg, the embryonic cell fates are **indeterminate**, the **blastopore** becomes the anus instead of the mouth, and the **coelom** forms from **enterocoelic pouches** of the primitive gut. The coelom is further divided into three separate coelomic compartments and a body region is associated with each: the anterior most **protocoel** and the **protosome**, the middle **mesocoel** and **mesosome**, and the **metacoel** and **metasome**. In the lophophorate animals, the mesosome becomes the **lophophore**, the metasome the main body cavity, and the prosome is usually reduced to a small flap covering the oral opening or it is missing. As you look at each of the different **deuterostome** phyla be sure to watch for the fates of each of these different coelomic spaces.

Lophophore

The lophophore consists of a ring of hollow, ciliated tentacles surrounding the mouth but not the anus located outside the ring. The mesocoel extends into each tentacle held erect by hydrostatic pressure and moved by muscles inside. The lophophore is used for suspension feeding, and lateral cilia on the sides of the tentacles pull water down into the center of the lophophore and out between the bases of the tentacles. As particles of food pass between the tentacles, they're trapped by cilia and passed toward the mouth. Different lophophorates modify this basic plan. In bryozoans, lateral cilia create the current; when they contact particulate food, these cilia pass it down and toward the frontal cilia that pass it back to the lateral cilia. They continue to bat the particle of food back and forth until it reaches the mouth. In the brachiopods, the lophophore is coiled, unmoveable, and housed inside the protective shell. Like the bryozoans the lateral cilia create the current; when they contact food, it is passed to the frontal cilia but only the frontal cilia pass it down the length of the tentacle to the ciliated food grove extending the length of the coiled lophophore to the mouth. Phoronids use mucus to trap particulate food, and in these animals, cilia pump water and a food-laden string of mucus toward the mouth. The lophophore is also an important respiratory surface in all three phyla, and in the smaller bryozoans and brachiopods the coelomic fluid is the main circulatory fluid. In the larger, burrowing phoronids blood vessels extend into the lophophore where it is oxygenated before flowing down the efferent vessel to the stomach at the distal end of the U-shaped gut where it absorbs nutrients before passing up the afferent vessel and back to the lophophore.

U-shaped gut

Sessile animals that live in burrows or shells by necessity have a U-shaped gut with the mouth and anus located close to each other. The trick is to position the anus so that wastes from it can't mistakenly enter the mouth. In lophophorate animals, the anus is located outside the lophophore ring, and wastes are flushed away from the animal as water passes between the tentacles.

✍ *U-shaped guts aren't unique in animals. What other animals have them?*

A Closer Look Inside *Digital Zoology* - Bryozoa and Brachiopoda

Marine bryozoan: Bugula

Bugula is a colonial marine bryozoan commonly found attached to rocks; seaweeds; and just about anything, including sponges and molluscs, in shallow coastal waters. Without the aid of a microscope it looks more like a plant than an animal, and you can see how bryozoans got their common name of moss animals. The colony is the **zooarium**, and each individual is a **zooecium** composed of soft parts, **polypide**, and the casing, **cystid**. The colony grows by dichotomous branching, and the oldest parts of the colony are at the base of the branches, the youngest at the top. In *Bugula* the cystid is made of **chitin** and transparent enough to be able to see the internal structures. The lophophore consists of fourteen tentacles surrounding the mouth, and it may be either extended or withdrawn through the aperture of the cystid by retractor muscles. The mouth opens into the pharynx, and the large saclike stomach is divided into cardiac and pyloric regions separated by a cecum. A short intestine leads away from the stomach and connects to the anus that opens outside the tentacular ring of the lophophore. If you look closely, you'll see a specialized zooecium that resemble a bird's head, the **avicularium**. The lower "beak" is moveable, and the two sets of muscles that open and close it can be seen through the wall of the avicularia.

✍ *What do the avicularia do?*

Freshwater bryozoans: **Pectinatella** *and* **Plumatella**

Pectinatella is common in slow-moving rivers, lakes, and ponds and can form large colonies several feet in diameter. The zooarium is gelatinous, and instead of individual cystids, the gelatinous mass is a common skeleton for the whole colony with individual animals sitting in it. The lophophore is horseshoe shaped with a double row of tentacles and the mouth at its base. The pharynx connects the mouth to the stomach with the cardiac and pyloric regions separated by the cecum. A short intestine connects the pyloric stomach to the anus outside the ring of tentacles. Look closely and you'll see the retractor muscles that pull the lophophore in and the **funiculus** with statoblasts on it. The funiculus provides a mechanism for individuals in the colony to share nutrients and is connected to the end of the **digestive cecum**. The **statoblast** is a resistant stage of the bryozoan that allows it to survive the adverse conditions associated with freshwater environments that may become anoxic, dry up, or freeze. *Plumatella* grows in branching colonies with a chitinous outer casing; internal structures are similar to those of *Pectinatella*.

 Protective stages like the statoblast are common in freshwater animals. Can you think of a few other examples?

Brachiopoda: *Lingula*

Lingula is one of the oldest known genera of animals, dating back to the Cambrian; today they're found living between the high- and low-tide lines in tropical and subtropical oceans. The long **pedicle** is hollow and anchors the animal in the soft sand and mud where it lives. The body is housed inside two bivalved shells secreted by an underlying **mantle**. The anterior edge of the shells is exposed above the substrate, and when it's open, cilia on the lophophore inside the mantle cavity create water currents that bring food into the mantle cavity. The inhalant flow occurs at the edges and exhalant at the center of the anterior opening protected by setal hairs. Once the shell is removed and the mantle folded back you can see the coiled lophophore with its two rows of tentacles. Once you remove the mantle that covers the viscera, the gonads, digestive gland, **metanephridium**, intestine, and abductor and adductor shell muscles are visible. To see the other structures you'll need to carefully remove the gonads. *Lingula* is an **articulate** brachiopod; it doesn't have any teeth to align the shells, and there is an additional set of muscles that control the lateral position of the shell.

 What are some of the similarities and differences between the bivalve shell of a brachiopod and a mollusc bivalve?

Cross-References to Other McGraw-Hill Zoology Titles

Integrated Principles of Zoology, 11th edition. C. P. Hickman, L. S. Roberts, and A. Larson. Chapter 22.

Animal Diversity, 3rd edition. C. P. Hickman, L. S. Roberts, and A. Larson. Chapter 13.

Zoology, 5th edition. S. A. Miller and J. P. Harley. Chapter 16.

Biology of the Invertebrates, 4th edition. J. Pechenik. Chapter 19.

Structures Checklist

Here are some of the structures that you should be able to easily find in *Digital Zoology* and the specimens that you will be looking at in your lab. After reading your lab handout, you might want to add more and, depending on the equipment available in your lab, you might see more. As you study the material, you might also want to make some notes on how some of these structures looked or include a drawing in your lab notes. (Structures indicated by * may be hard to see.)

Brachiopoda - Lingula

External anatomy
- [ ] Dorsal shell (brachial shell)
- [ ] Gelatinous cuticle
- [ ] Growth rings
- [ ] Pedicile
- [ ] Setal hairs
- [ ] Shell
- [ ] Shell apex
- [ ] Ventral shell (pedicile shell)
- [ ]
- [ ]

- [ ]
- [ ]
- [ ]
- [ ]

Internal anatomy
- [ ] Abductor (diductor) muscle
- [ ] Adductor muscle
- [ ] Anus
- [ ] Branchium of the lophophore
- [ ] Coelomic cavity
- [ ] Digestive gland

- [ ] Dorsoventral transverse muscle
- [ ] External oblique muscle
- [ ] Gonad
- [ ] Ilioparietal ligament
- [ ] Intestine
- [ ] Lophophore tentacles
- [ ] Mantle
- [ ] Mantle canals
- [ ] Mantle cavity
- [ ] Median oblique muscle

- [ ] Metanephridium
- [ ] Mouth
- [ ] Stomach
- [ ]
- [ ]

Additional structures
- [ ]
- [ ]
- [ ]
- [ ]

Freshwater Bryozoans - Plumatella

- [ ] Anus
- [ ] Cardiac region of stomach
- [ ] Cecum
- [ ] Cystid

- [ ] Epistome
- [ ] Funiculus
- [ ] Lophohore retractor*
- [ ] Lophophore

- [ ] Pharynx
- [ ] Polypide
- [ ] Pyloric region of stomach
- [ ] Statoblast

- [ ] Zoecium
- [ ]
- [ ]
- [ ]
- [ ]

Freshwater Bryozoans - Pectinatella

- ☐ Anus
- ☐ Calyx
- ☐ Cardiac region of stomach
- ☐ Cecum
- ☐ Cystid
- ☐ Developing embryo

- ☐ Epistome
- ☐ Funiculus
- ☐ Lophophore
- ☐ Lophophore retractor
- ☐ Ovary*
- ☐ Pharynx

- ☐ Polypide
- ☐ Pyloric region of stomach
- ☐ Statoblast
- ☐ Testis
- ☐ Zoecium
- ☐

- ☐
- ☐

Marine Bryozoans - Bugula

External anatomy

- ☐ Anus
- ☐ Avicularium
- ☐ Cardiac region of stomach
- ☐ Cecum
- ☐ Cystid
- ☐ Intestine

- ☐ Lophophore
- ☐ Lophophore retractor muscle
- ☐ Mouth
- ☐ Pharynx
- ☐ Polypide
- ☐ Pyloric region of stomach

- ☐ Zoecium
- ☐
- ☐

Avicularium

- ☐ Abductor muscle
- ☐ Adductor muscle
- ☐ Beak

- ☐ Mandible
- ☐ Pedicle
- ☐ Vestigal polypide
- ☐
- ☐

Crossword Puzzle–Bryozoa and Brachiopoda

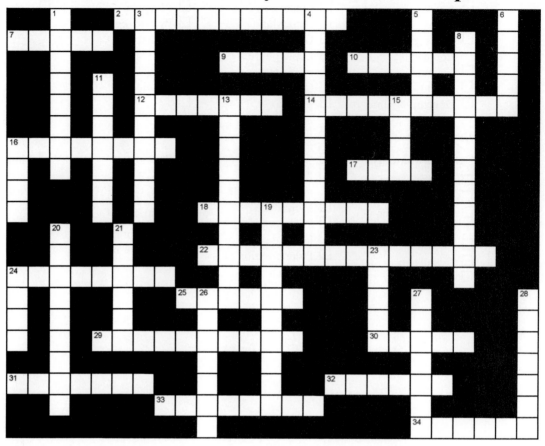

Puzzle solution. An interactive web-based version of this puzzle and its solution are available on the *Digital Zoology* website www.digitalzoology.com/students. With the interactive puzzle you can check to see if individual words or the whole puzzle is correct and get hints for single letters.

Across

2 This unusual respiratory pigment isn't in a brachiopod's blood; it's in the coelomic fluid. (11)

7 An individual bryozoan. (5)

9 These cover the surface of a bryozoan's tentacles. (5)

10 Like a mollusc, this secretes a brachiopod's shell. (6)

12 This attaches a brachiopod to the substrate. (7)

14 Like other lophophorates, a bryozoan coelom is described as this. (10)

16 These muscles close the brachiopods shell. (8)

17 Bryozoans get their common name from these plants, which they look like. (4)

18 The inner layer of a brachiopod shell. (9)

22 Excretory structure found in a brachiopod. (14)

24 Unlike the other lophophorate phyla, bryozoans exhibit this type of lifestyle. (8)

25 Brachiopods are found in this part of the marine environment. (6)

29 Every bryozoan colony starts from one of these. (10)

30 There are a lot of fossil bryozoans because of this structure. (5)

31 Another name for a bryozoan colony. (7)

32 The outer casing and living part of the bryozoan that produces it. (6)

33 Unlike molluscs that use the elasticity of the hinge ligament to open their shells, brachiopods use this muscle to open theirs. (8)

34 Most bryozoan species live in this environment. (6)

Down

1 The part of a bryozoan that includes the lophophore and retractable parts of the body. (8)

3 Another formal name often used for the phylum Bryozoa. (10)

4 If a tooth and socket hinge is missing in a brachiopod it is referred to as being of this type. (12)

5 Position of the periostracum layer in a brachiopod shell. (5)

6 Found around the edge of the opening to the mantle cavity, these keep large particles out of the cavity. (5)

8 The outer layer of the brachiopod shell. (12)

11 These pull a bryozoan's tentacles back inside its casing. (7)

13 Brachiopods and this mollusc class are good examples of this type of evolution. (10)

15 Articulate brachiopods are missing this part of the digestive tract; articulate species still have it. (4)

16 A brachiopod's lophophore is found in these. (4)

19 These help freshwater bryozoans get through harsh environmental conditions--winter, for example. (10)

20 A bryozoan's body symmetry. (9)

21 This complex carbohydrate is part of a brachiopod shell. (6)

23 You can tell the age of a brachiopod by these on its shell. (5)

24 Arrangement of the tentacular arms in a brachiopod. (4)

26 The position of the bryozoan anus relative to the lophophore. (7)

27 The outer casing of a bryozoan. (7)

28 Describes a brachiopod shell. (7)

Self-Test - Bryozoa and Brachipoda

Use the following labels to identify the photographs. You may have to use a label more than once, and some labels may not be appropriate for the photographs. Answers are available on the *Digital Zoology* website at www.digitalzoology.com/students. Be sure to try the interactive drag-and-drop quizzes t on the *Digital Zoology* CD-ROM.

Specimens	Labels		
A- Brachiopoda - *Lingula*	1) Adductor muscle	7) Developing embryo	13) Metanephridium
	2) Anus	8) Digestive gland	14) Polypide
B- Bryozoa - *Plumatella*	3) Budding polypide	9) Ilioparietal ligament	15) Pyloric stomach
	4) Cardiac stomach	10) Intestine	16) Statoblast
C- Bryozoa - *Pectinatella*	5) Cecum	11) Lophophore	17) Stomach
	6) Cystid	12) Lophophore retractor muscle	

ECHINODERMATA

Inside *Digital Zoology*

As you explore the Echinodermata on the *Digital Zoology* CD-ROM or on the web at www.digitalzoology.com, don't miss these learning tools:

- Photos of dissected specimens of the preserved sea star and sea urchin and a look at brittle stars, sand dollars, and some echinoderm fossils.
- Drag-and-drop quizzes on the key structures of the sea urchin and sea star.
- Interactive cladogram showing the major events that gave rise to the five living echinoderm classes. A summary of the key characteristics of each class are combined with an interactive glossary of terms.

Defining Differences

Some of the differences described in the following sections appear for the first time in the Echinodermata, and they define the phylum. Others are important for understanding how the animals in the phylum function. Whichever is the case, you'll want to watch for examples of each in the various echinoderms that you will be examining.

- Pentaradiate symmetry with unique
- tube feet and water vascular system, an
- endoskeleton, and protection by
- pedicellaria.

Pentaradiate symmetry

In a mixed basket of invertebrates you would probably have no trouble picking out most of the echinoderms. Not because you knew their names, but because of their **radial symmetry,** most often based on fives, called **pentaradiate symmetry**. At one time biologists put echinoderms and cnidarians into a single **taxon**–the Radiata. The assumption was that the cnidarians had the **diploblastic** radial plan that ultimately evolved into the **triploblastic** echinoderms. It made sense until investigators found that echinoderm larvae were actually **bilaterally symmetric** and underwent **metamorphosis** to become radially symmetric.

✍ *What are the different types of symmetry that animals can have?*

There are distinct advantages to being a bilaterally symmetric animal, and most animal body plans have this type of symmetry. But, for some reason, the echinoderms chose to do things differently and returned to radial symmetry. There must have been some advantage for doing that, and to find out what it was we have to take a look at the origins of the phylum.

Like a variety of other ancient animals that existed at that time, sessile sponges and cnidarians for example, ancestral echinoderms were **suspension feeders** who trapped and captured food from the water that surrounded them. Echinoderms developed a unique way to catch their food using the tube feet that lined each of the arms that surrounded their central mouth. Once caught, the food was moved to the mouth by a combination of tube feet and a ciliated groove that ran the length of the arm. A new way of suspension feeding was developed and along with it the beginnings of the echinoderms.

✍ *How do sponges and cnidarians trap and capture their food?*

It worked well for them, and when they gave up their sessile existence to move across the substrate, echinoderms kept their basic radial symmetry. As you'll see when you look at echinoderms in your lab, or in *Digital Zoology*, this has been modified in a variety of ways. There are even some echinoderms that show signs of returning to bilateral symmetry!

Tube feet and the water vascular system

Like all deuterostomes, echinoderms have a **tripartite coelom.** Part of the **mesocoel** has been used to form the unique **water vascular system,** which includes the tube feet that extend from the oral side of an echinoderm. Each **tube foot** is connected to the rest of the water vascular system and ultimately to external sea water through the **madreporite** found on the aboral side of the animal. Individually each tube foot can be hydrostatically isolated from the rest of the system allowing it to function as a small **hydrostatic skeleton**. A combination of muscle contractions in the wall of the tube foot and the suction cup action of the tip, allows each of the tube feet to pull the echinoderm along. It might seem inefficient, but with thousands of tube feet working together, echinoderms are able to move, although slowly. Few echinoderms display the ancestral use of the tube feet in feeding. Modern echinoderms have their aboral surface facing down, and the tube feet are used for locomotion rather than feeding.

✍ *What are the different parts of the water vascular system, and how are they connected to each other?*

Endoskeleton

Echinoderms get their name from their spiny skin and their spines, made of **ossicles** and embedded in the **mesoderm,** creating the **endoskeleton**. What makes an echinoderm's endoskeleton unique is how it's made of calcareous ossicles that start as single calcite spicules. Gradually, as more calcite is deposited, an ossicle takes on its final shape. They aren't solid; instead they're a hollow mesh of crystallized calcite filled with the living tissue. Their different shapes are key characteristics that echinoderm taxonomists use to identify different classes and species in the phylum. Ossicles may be small or fuse to form large skeletal plates.

Pedicellaria

The pincerlike **pedicellaria** found on the surface of some echinoderms are good examples of highly modified ossicles. It was once thought that these were some sort of **ectoparasite**, especially because they have their own musculature and react to their environment without any connection to the rest of the animal. Their role is debatable. Some prevent other animals from sitting on the echinoderm, suggesting a defensive role. There are even pedicellaria that inject toxic poisons to help protect the echinoderm. Still others seem to be responsible for keeping the surface of the animal clean and clear of any debris that settles on the animal. After all, these aren't the fastest animals in the oceans and anything falling through the water column has the potential to land on them.

A Closer Look Inside *Digital Zoology* - Echinodermata

Sea star

The sea star, *Asterias*, is a good introduction to the general features of the phylum Echinodermata. Try rubbing your fingers across the aboral surface of the specimen. You'll feel the tiny spines and understand how the phylum got its name. Even larger ambulacral spines can be found on the oral surface where they protect the delicate tube feet found in the groove. The tube feet and the madreporite on the aboral surface are the only parts of the water vascular system that you can see externally. You can see the rest when you dissect the specimen.

Internally, each arm contains the digestive glands and gonads, surrounded by a spacious perivisceral body cavity. The proximity of the digestive gland to the gonad is handy, allowing ingested nutrients to pass quickly to the gonads, transported by the fluids of the **perivisceral cavity**. Fluid in the perivisceral cavity is an important circulatory fluid that not only carries nutrients to the gonads and other tissues, but it also absorbs oxygen through fingerlike projections of the wall of the cavity, called the **dermal branchia**, which extend through the body wall to the outside.

The dermal branchia aren't the only places where gas exchange occurs. The extensive surface of all of those tube feet is equally important, and oxygen diffuses across them to the fluid that fills the water vascular cavity. Exchange with the perivisceral fluid occurs across the ampullae of the feet, which extend into the main body cavity. The water vascular system and perivisceral body cavity are lined with cilia that beat making sure that the contents of both are always well mixed.

Just as oxygen can diffuse in, so too can nitrogenous wasters diffuse out, and both dermal branchia and tube feet are important surfaces for removing these wastes. As you can see, the importance of tube feet in how the sea star functions goes beyond locomotion.

Inside the central disc, the first of two stomachs, the **cardiac stomach**, can turn inside out when the animals feed. As you might imagine, being a predatory echinoderm is no easy task when most of the animals around you move considerably faster. But there are a few animals that are potential prey, and one of them is bivalve clams. The closed shells of these molluscs are usually effective protection from most predators but the sea star is able to take advantage of one of the minor flaws in a clam's seemingly impervious armor. The shells rarely close perfectly, and there are small gaps, or spaces, along the edges where the two shells meet. The sea star surrounds the clam, holding the edges of the shell against its mouth. The sea star then everts the membranous cardiac stomach squeezing it through these small openings in the clam's closed shell. Digestive enzymes released into the clam start to break down the clam's tissues and cilia lining the sea star's digestive tract transport the nutrients into the second stomach, the **pyloric stomach**, and out to the pyloric ceca in each arm.

Sea urchin and sand dollar

The sea urchin and sand dollar have modified their original body plans but in two different ways–by either forming a hollow globe (regular) or a flattened solid disc (irregular).

Sea urchins

With a sea urchin you have to look closely to see what remains of the five ambulacral regions. The surface of the spherical **test** is alternating ambulacral and interambulacral regions with the long tube feet extending past the spines that cover the animal. Both the tube feet and the spines are important for locomotion. The ball-and-socket joint that anchors each spine to the surface of the test is one of the rare occurrences of this type of joint in invertebrates.

Protruding from the center of the aboral surface, you can see the tips of the five teeth of Aristotle's lantern formed by the same process that makes the endoskeleton. In this case, the ossicles form complex pyramid-shaped plates, each with an interlocking tooth. Musculature attached to the lantern allow the plates to move relative to each other, and the entire lantern can open, close, and rock back and forth as the urchin feeds. Surrounding the mouth are a variety of different types of pedicellaria, many of which are stalked.

Most urchins are herbivores, and one of things you'll notice when you open your urchin is the size of the intestine that winds its way through the test. It starts in one direction, then reverses direction as it makes its way to the anus at the top of the aboral surface. Like any other animal that has adopted an herbivorous existence, their gut is extensively modified to allow for processing large quantities of this nutrient-poor food, which explains its length. There is another unique modification that you'll see in the aboral part of the intestine. A **siphon** takes the water from the ingested meal and shunts it to a more distant part of the gut. This way the urchin is able to concentrate the food it consumes. The perivisceral cavity of the sea urchin is large, and suspended from the aboral surface are the five pairs of gonads.

 Why are plants considered a nutrient-poor food?

Sand dollars

Sand dollars are specialists at burrowing, and the long spines found in the sea urchins are considerably smaller in these animals. The five **petaloids** of the ambulacral regions appear on the aboral surface of the animal, and in sand dollars the tube feet are used primarily for gas exchange rather than locomotion. Sand dollars show signs of the return of bilateral symmetry. The oral opening is often positioned at one end of the disk rather than at the center, making it hard to define oral and aboral surfaces. The openings in the test are **lunnules,** and they also are examples of how sand dollars are returning to bilateral symmetry. The function of the lunnules is not certain. They may either stabilize the animals against water flow or may channel the food to the mouth of the animal.

Brittle stars and basket stars

Brittle stars get their name from their delicate arms, which easily break when they are handled. In these animals, ossicles in each of the arms form interlocking vertebrae, which can bend against each other in much the same way that the vertebral column does in vertebrates. Lateral extensions of the ossicles prevent individual vertebrae from slipping past each other, and bands of muscles that run the length of the arm allow it to easily bend and twist. Tube feet are no longer involved in locomotion. Instead, serpentine movements of the flexible arms are how these echinoderms move. It's a modification that makes them the most mobile in the phylum. If you look on the aboral surface of the arms, you won't find any tube feet; you won't even find an ambulacral grove. The central disc in the brittle star is distinct, and if you look at the aboral surface, you'll see the bursal slits, the openings to where the gonads are located.

The ancient history of the phylum can be seen in the fossils of the basket stars and some of their relatives. You can see the characteristics of the phylum, including the branched arms and their **pinnules**, which increased the surface area for capturing food.

Holothuroidea: the sea cucumber

Unlike other echinoderms, sea cucumbers have a fleshy body with calcareous ossicles separated from each other and embedded in the thick body wall. The mouth is neither pointing up nor down; these echinoderms have elongated the body along the oral aboral axis and lie on their sides. It's always the same side in contact with the substrate, and if you look closely at the tube feet you'll see that those underneath are longer than those on the opposite side. The other noticeable external feature is the bushy tentacles surrounding the mouth. These are modified tube feet that use mucus to trap food ingested by pushing the tentacles into the mouth. The pharynx scrapes the tentacles clean and applies a fresh coat of mucus to the tentacles before they're extended. You can watch a sea cucumber feeding in the video material available in *Digital Zoology*. The cloacal opening is at the posterior end and is the shared opening of the digestive tract and respiratory trees. When you cut the animal open you'll see that the large coelom is filled with two large structures, gonads and gut. The single gonad looks like a giant string mop, and its genital opening is located near the base of the tentacles. The digestive tract makes a giant loop through the body cavity before emptying into the cloaca. Paired respiratory trees are used for gas exchange, and seawater is pumped in and out through the cloacal opening. The madreporite of the water vascular system opens into the coelomic cavity rather than the surrounding seawater--locate it and other parts of the water vascular system. Five retractor muscles are arranged around the body wall and coincide with the ambulacral grooves where you'll see the ampullae of the tube feet extending into the coelomic cavity.

Cross-References to Other McGraw-Hill Zoology Titles

Integrated Principles of Zoology, 11th edition. C. P. Hickman, L. S. Roberts, and A. Larson. Chapter 23.

Animal Diversity, 3rd edition. C. P. Hickman, L. S. Roberts, and A. Larson. Chapter 14.

Zoology, 5th edition. S. A. Miller and J. P. Harley. Chapter 16.

Biology of the Invertebrates, 4th edition. J. Pechenik. Chapter 20.

Laboratory Studies in Integrated Principles of Zoology, 10th edition. C. P. Hickman, F. Hickman, and L. Kats. Chapter 16.

General Zoology Laboratory Guide, 13th edition. C. Lytle. Chapter 14.

Structures Checklist

Here are some of the structures that you should be able to easily find in *Digital Zoology* and the specimens that you will be looking at in your lab. After reading your lab handout, you might want to add more and, depending on the equipment available in your lab, you might see more. As you study the material, you might also want to make some notes on how some of these structures looked or include a drawing in your lab notes. (Structures indicated by * may be hard to see.)

Sea star

External anatomy

- [ ] Aboral surface
- [ ] Ambulacral groove
- [ ] Ambulacral spines
- [ ] Anus*
- [ ] Arms - bivium
- [ ] Arms - trivium
- [ ] Central disk
- [ ] Dermal branchia
- [ ] Eye spot*
- [ ] Madreporite
- [ ] Mouth
- [ ] Oral surface
- [ ] Pedicellaria
- [ ] Spines
- [ ] Tube feet
- [ ]
- [ ]
- [ ]
- [ ]
- [ ]
- [ ]
- [ ]
- [ ]

Internal anatomy

- [ ] Ambulacral ossicles
- [ ] Ampullae of the tube feet
- [ ] Cardiac stomach
- [ ] Genital pore*
- [ ] Gonads
- [ ] Lateral canal
- [ ] Pyloric (digestive) ceca
- [ ] Pyloric duct
- [ ] Pyloric stomach
- [ ] Radial canal
- [ ] Ring canal *
- [ ] Stone canal
- [ ]
- [ ]
- [ ]

Additional structures

- [ ]
- [ ]
- [ ]

Sea urchin

External anatomy

- [ ] Aboral surface
- [ ] Ambulacral region
- [ ] Anus
- [ ] Buccal podia
- [ ] Genital plates
- [ ] Interambulacral region
- [ ] Madreporite plate
- [ ] Mouth
- [ ] Oral surface
- [ ] Pedicellaria
- [ ] Peristomial gills
- [ ] Spines
- [ ] Teeth of Aristotle's lantern
- [ ] Tube feet
- [ ]
- [ ]
- [ ]
- [ ]

Internal anatomy

- [ ] Aboral intestine
- [ ] Ambulacral plates
- [ ] Ambulacral region
- [ ] Ampullae of the tube feet
- [ ] Aristotle's lantern
- [ ] Compass
- [ ] Dental sac
- [ ] Epiphysis
- [ ] Esophagus
- [ ] Gonads
- [ ] Interambulacral region
- [ ] Lantern protractors
- [ ] Lantern retractor
- [ ] Oral intestine (stomach)
- [ ] Pyramid
- [ ] Rectum
- [ ] Ring canal*
- [ ] Radial canal*
- [ ] Siphon
- [ ] Tooth
- [ ]
- [ ]
- [ ]
- [ ]
- [ ]
- [ ]

Additional structures

- [ ]
- [ ]

Sand dollar

External anatomy

- [ ] Ambulacral region
- [ ] Food groove
- [ ] Genital plates
- [ ] Genital pores
- [ ] Interambulacral region
- [ ] Lunnules
- [ ] Madreporite
- [ ] Oral surface
- [ ] Position of the anus
- [ ] Position of the mouth
- [ ] Spines
- [ ]
- [ ]
- [ ]

Additional structures

- [ ]
- [ ]
- [ ]
- [ ]

Brittle star

External anatomy

- ☐ Aboral surface
- ☐ Arm spines
- ☐ Buccal tube feet
- ☐ Bursal slits
- ☐ Central disk

- ☐ Jaws
- ☐ Mouth
- ☐ Oral surface
- ☐ Radial shield
- ☐
- ☐

☐

Additional structures

- ☐
- ☐
- ☐

- ☐
- ☐
- ☐

Sea cucumber

External anatomy

- ☐ Ambulacral region
- ☐ Cloacal opening
- ☐ Dorsal surface
- ☐ Genital pore
- ☐ Interambulacral region
- ☐ Mouth
- ☐ Podia of tube feet
- ☐ Tentacles
- ☐ Ventral surface
- ☐
- ☐

- ☐
- ☐
- ☐
- ☐

Internal anatomy

- ☐ Ampullae of tube feet
- ☐ Ascending intestine
- ☐ Calcerous ring
- ☐ Cloacal dilators
- ☐ Descending intestine

- ☐ Dorsal mesentery
- ☐ Esophagus
- ☐ Gonad
- ☐ Gonadal duct
- ☐ Introvert retractor muscle
- ☐ Longitudinal muscle
- ☐ Madreporite
- ☐ Polian vesicle
- ☐ Rectum
- ☐ Respiratory tree
- ☐ Stomach

- ☐ Stone canal
- ☐

Additional structures

- ☐
- ☐
- ☐
- ☐
- ☐
- ☐
- ☐
- ☐

Crossword Puzzle–Echinodermata

Puzzle solution. An interactive web-based version of this puzzle and its solution are available on the *Digital Zoology* website www.digitalzoology.com/students. With the interactive puzzle you can check to see if individual words or the whole puzzle is correct and get hints for single letters.

Across

1 Whose lantern was used to name the unique sea urchin feeding apparatus? (9)

4 This part of the tube foot extends into the body cavity of an echinoderm. (7)

7 Cardiac and pyloric describe these two structures in a sea star. (8)

9 A larval echinoderm has this type of symmetry. (9)

14 Most echinoderms have this number of arms. (4)

15 In brittle stars the ossicles in the arms resemble these. (9)

17 This describes the body cavity of a sea urchin compared to a sand dollar. (8)

21 In an echinoderm, the ring canal connects to this canal before connecting to the tube feet. (6)

23 Water enters the water vascular system through openings in this structure. (11)

24 This canal connects the madreporite to the ring canal in an echinoderm. (5)

25 These holes let water pass from the upper to the lower surface of a sand dollar. (8)

28 In sea stars, gas exchange occurs in the tube feet and these. (6,8)

30 Like all deuterostomes, echinoderms have this type of divided coelom. (10)

33 The ancestral echinoderms tube feet were originally used for this. (7)

35 In an echinoderm the tip of the tube foot uses this to help it attach to the substrate. (7)

36 The name of the body cavity that surrounds the internal organs in an echinoderm. (12)

37 The arms of a sea star are connected to this part of the animal. (7,4)

38 These defensive structures cover the surface of some echinoderms. (12)

Down

2 This term describes the circulatory and nervous systems in echinoderms. (7)

3 The number of digestive ceca in each arm of a starfish. (3)

5 Another name for the digestive ceca found in the arms of a sea star; it's the same as the stomach to which they are attached. (7)

6 The main locomotory structures used by echinoderms. (4,4)

8 Individually each tube foot is this type of skeleton; it's important for understanding how echinoderms move. (11)

10 The radial canals of an echinoderm's water vascular system are connected to these canals, and then to the tube feet. (7)

11 In echinoderms, the blastopore ultimately forms this structure. (4)

12 Echinoderms have this special type of radial symmetry. (12)

13 The water vascular system is formed from this part of the coelom. (8)

16 An echinoderm's skeleton is this type. (12)

18 Without bilateral symmetry, we refer to the main surface opposite the mouth in echinoderms as this side of the animal. (6)

19 Relative to the surround sea water, the water in the water vascular system is this. (8)

20 The appearance of radial symmetry in the echinoderms is referred to as this type of evolutionary event. (9)

22 Absence of this system has limited echinoderms to the marine environment. (9)

26 Ancestrally, the mouth of an echinoderm faced in this direction relative to the substrate to which it was attached. (2)

27 Urchins concentrate their ingested meal by removing the water from it using this modification of the digestive tract. (6)

29 Echinoderms get their name from these structures found on their surface. (6)

31 When present these vesicles act as reservoirs that store water for the water vascular system. (6)

32 Sea urchin shells are also called this. (5)

34 These keep the fluids on both echinoderm body cavities moving. (5)

Self-Test - Echinodermata

Use the following labels to identify the photographs. You may have to use a label more than once, and some labels may not be appropriate for the photographs. Answers are available on the *Digital Zoology* website at www.digitalzoology.com/students. Be sure to try the interactive drag-and-drop quizzes on the *Digital Zoology* CD-ROM.

Specimens

A- Sea star
B- Sea cucumber
C- Sea urchin

Labels

1) Aboral intestine
2) Ambulacral ridge
3) Aristotle's lantern
4) Ascending intestine
5) Auricle
6) Cloaca

7) Cloacal dilators
8) Compass
9) Dental sac
10) Descending intestine
11) Epiphysis
12) Esophagus
13) Gonad

14) Introvert retractor
15) Lantern protractors
16) Lantern retractors
17) Oral intestine
18) Polian vesicle
19) Pyloric cecum
20) Pyloric duct
21) Respiratory tree

HEMICHORDATA

Inside *Digital Zoology*

As you explore Hemichordata on the *Digital Zoology* CD-ROM or on the web at www.digitalzoology.com, don't miss these learning tools:

○ Photos of preserved specimens and sagittal sections of *Balanoglossus*, the acorn worm.
○ Drag-and-drop quizzes on the key structures of *Balanoglossus*.
○ Interactive cladogram showing the major events that gave rise to the main deuterostome groups. A summary of the key characteristics of each of the deuterostome phyla and subphyla of chordates and hemichordates are combined with an interactive glossary of terms.

Defining Differences

Hemichordates include burrowing animals, such as the acorn worm, Enteropneusta, and the sessile colonial forms, such as the Pterobranchia. As a taxon they sit somewhere between the Echinodermata and Chordata in the evolution of **deuterostome** animals, but who they are most closely related to is still not clear. Not everyone agrees that the hemichordates should be a phylum, and there is increasing evidence that they are **polyphyletic** and that the two classes should be elevated to phyla. In this scenario the Enteropneusta would be more closely related to the chordates, the Pterobranchia more closely related to echinoderms. We take a look at Enteropneusta in *Digital Zoology*.

Some of the features in the following list appear for the first time in the Hemichordata, and they define the phylum. Other characters in the list we see for the first time in animals and are important for understanding how the organisms in the phylum function. Whichever is the case, you'll want to watch for examples of these in the specimens you'll be examining.

○ Body with a proboscis, collar, and trunk;
○ stomochord; and
○ glomerulus.

Proboscis, collar, and trunk

Deuterostomes have a **tripartite coelom** with an anterior **protocoel**, middle **mesocoel**, and posterior **metacoel**. In an enteropneust these correspond to the proboscis, collar, and trunk, respectively. Enteropneusts feed on either the substrate they burrow in or, if they live in permanent tubes, particulate food suspended in the water around them. Food is trapped in mucus covering the proboscis and cilia on the surface move it toward the back of the **proboscis** where the preoral **ciliary organ** either rejects or accepts the particles as food. Rejected food passes over the surface of the collar while acceptable material is concentrated in a mucus string that moves into the mouth under the anterior, ventral edge of the collar. Cilia lining the gut help move the mucus string. Only the mucus string enters the digestive system; any water that's pumped into the mouth by the cilia passes out through the gill pores. The collar contains two coelomic cavities and musculature to form a hydrostatic skeleton used to anchor the worm in its burrow or tube. The proboscis is also used in locomotion and changes shape as it pulls the acorn worm across the substrate or, in combination with the collar, to move the animal in its tube or burrow. The trunk is divided into regions: a **branchial region** where the gill pores open, a **genital region** where the gonads are located, and a **hepatic region** where the intestine is modified with ceca.

Stomochord

Hemichordates get their name from having most, but not all, of the chordate characters: **pharyngeal gill slits**, a **dorsal hollow nerve cord** in some species, and what appears to be a **notochord**. In chordates the notochord forms along the dorsal surface of the entire gut and completely separates to become the unique solid, skeletal element of the phylum. Although the **stomochord** is also derived from the primitive gut, that's where the similarities end. The stomochord, or **buccal diverticulum** as it's also called, forms from the mesocoel, extends as a hollow pocket into the proboscis with an opening into the gut, and acts as a hydrostatic skeleton supporting the proboscis. The stomochord is analogous, not homologous, to the chordate notochord. The other missing characteristic is a tail; the anus is on the posterior tip of the body.

Glomerulus

The **glomerulus** is a unique excretory organ found only in hemichordates. Until recently it wasn't clear how it worked or even if it was excretory. It now appears that when blood is forced into the central sinus by the **heart vesicle** there is sufficient pressure to force fluid from the blood across the glomerulus and into the protocoel. Essential nutrients and ions are recovered from the coelomic fluid of the proboscis by cells lining the cavity, and urine is released through the proboscis pore.

A Closer Look Inside *Digital Zoology* - Hemichordata

Balanoglossus: *The acorn worm*

It is easy to see the three main body regions in the acorn worm: proboscis, collar, and trunk. The proboscis is preoral, muscular, and attached by a short stalk to the collar that overlaps the proboscis and trunk. As you handle the trunk you'll notice that it seems flimsy; that's because there are no muscles in the trunk. The anterior part of the trunk has been modified into genital wings surrounding the branchial region with gill pores and the genital region. Fingerlike projections of the body wall identify the hepatic region behind where the genital wings fuse.

Sagittal sections of the acorn worm give you a chance to see some of the structures characteristic of the group and others they share with the chordates. It is easy to identify the proboscis, its musculature, and protocoel inside. Behind the proboscis is the collar, and the cavity of the metacoel is another feature that's not hard to find. The proboscis complex is a mass of tissue between the proboscis and collar and includes the stomochord, heart vesicle, and glomerulus. If you're having trouble distinguishing one from the other, take a look at the photos available in *Digital Zoology*. As blood in the dorsal vessel flows toward the anterior end of the animal, it combines with blood from the proboscis and pools in the central sinus located between the stomochord and the heart vesicle. It is called a heart vesicle because blood doesn't pass through it; instead, the vesicle pushes against the central blood sinus to squeeze the blood through the sinus in the same way that a peristaltic pump squeezes fluid through tubing. We've already taken a look at the glomerulus and stomochord.

Cross-References to Other McGraw-Hill Zoology Titles

Integrated Principles of Zoology, 11th edition. C. P. Hickman, L. S. Roberts, and A. Larson. Chapter 24.

Animal Diversity, 3rd edition. C. P. Hickman, L. S. Roberts, and A. Larson. Chapter 14.

Zoology, 5th edition. S. A. Miller and J. P. Harley. Chapter 17.

Biology of the Invertebrates, 4th edition. J. Pechenik. Chapter 21.

Structures Checklist

Here are some of the structures that you should be able to easily find in *Digital Zoology* and the specimens that you will be looking at in your lab. After reading your lab handout, you might want to add more and, depending on the equipment available in your lab, you might see more. As you study the material, you might also want to make some notes on how some of these structures looked or include a drawing in your lab notes. (Structures indicated by * may be hard to see.)

The acorn worm

External anatomy

- ☐ Abdominal region of trunk
- ☐ Anus
- ☐ Branchial region of trunk
- ☐ Collar
- ☐ Genital region of trunk
- ☐ Genital wings
- ☐ Gill pores
- ☐ Hepatic region of trunk

- ☐ Mouth*
- ☐ Proboscis
- ☐ Proboscis stalk
- ☐ Trunk
- ☐
- ☐

Sagittal section

- ☐ Central blood sinus
- ☐ Circular proboscis musculature
- ☐ Collar
- ☐ Collar coelom
- ☐ Dorsal nerve cord

- ☐ Dorsal vessel
- ☐ Glomerulus
- ☐ Heart blood sinus
- ☐ Heart vesicle
- ☐ Mouth
- ☐ Pharynx
- ☐ Proboscis coelom (protocoel)
- ☐ Proboscis musculature
- ☐ Stomochord (buccal diverticulum)
- ☐ Ventral vessel

- ☐
- ☐
- ☐

Additional structures

- ☐
- ☐
- ☐
- ☐

Crossword Puzzle – Hemichordata

Puzzle solution. An interactive web-based version of this puzzle and its solution are available on the *Digital Zoology* website www.digitalzoology.com/students. With the interactive puzzle you can check to see if individual words or the whole puzzle is correct and get hints for single letters.

Across

2 In acorn worms the protocoel forms this structure. (9)

4 In hemichordates the stomochord was mistakenly thought to be this structure and the reason that they were originally included as a subphylum of the Chordata. (9)

8 Hemicordates are missing a notochord and this chordate character; they have the other two. (4)

9 In a hemichordate acorn worm, this diverticulum of the gut was originally thought to be a notochord. (10)

12 These fingerlike projections of the body wall identify this part of the an acorn worm's trunk. (7)

13 The direction in which blood flows in the dorsal blood vessel of an acorn worm. (8)

14 Even though they have pharyngeal slits and pores, acorn worms don't have these. (5)

16 On the inside of an acorn worm's pharynx, these are U-shaped and connected to pores on the outside. (9)

17 Enteropneust hemichordates resemble this type of animal. (5)

18 The collar of a hemichordate acorn worm is formed from this coelomic space. (8)

20 Hemichordates may provide some clues on the origins of this animal phylum. (8)

22 Contraction of these muscles in the collar help anchor an acorn worm in its burrow. (12)

23 Muscles in this pump blood into the glomerulus of an acorn worm. (12)

Down

1 In the trunk of a hemichordate there is a nerve cord on the dorsal and ventral surface; in the collar it's only on this side. (6)

3 This creates the blood filtrate found inside the glomerulus of an acorn worm. (5,8)

4 Like cnidarians and echinoderms, this best describes the organization of the nervous system in hemichordates. (3)

5 Like all deuterostomes, hemichordates have this type of coelom. (10)

6 You won't find any of these in the trunk of an acorn worm. (7)

7 An alternate name for the stomochord is this type of diverticulum. (6)

10 The preoral ciliary organ mixes food particles with this before passing them into an acorn worm's mouth. (5)

11 The presence of gill pores defines this part of an acorn worm's trunk. (9)

15 The presumed excretory structure of a hemichordate acorn worm. (10)

18 The only environment where you'll find a hemichordate. (6)

19 These move food trapped on the surface of an acorn worm's proboscis into the gut. (5)

21 The stomochord opens into this. (3)

Self-Test – Hemichordata

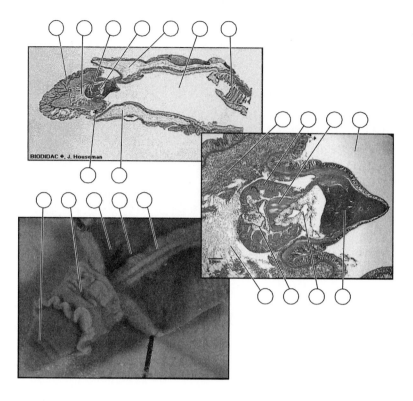

BIODIDAC ♦, J. Houseman

Use the following labels to identify the photographs. You may have to use a label more than once, and some labels may not be appropriate for the photographs. Answers are available on the *Digital Zoology* website at www.digitalzoology.com/students. Be sure to try the interactive drag-and-drop quizzes on the *Digital Zoology* CD-ROM.

Specimen

Balanoglossus

Labels

1) Branchial body region
2) Central blood sinus
3) Collar
4) Genital wings
5) Gill pores
6) Glomerulus
7) Heart blood sinus
8) Heart vesicle
9) Pharyngeal slits
10) Pharynx
11) Proboscis
12) Proboscis coelom (protocoel)
13) Proboscis musculature
14) Stomochord (buccal diverticulum)

Urochordata and Cephalochordata

Inside *Digital Zoology*

As you explore the Urochordata and Cephalochordata on the *Digital Zoology* CD-ROM or on the web at www.digitalzoology.com, don't miss these learning tools:

○ Photos of preserved specimens and cross sections of the hemichordate *Balanoglossus*, whole mounts and cross sections of larval and adult urochordates, a cephalochordate, the lancelet *Branchiostoma*, and the ammocoete larva of the lamprey all help to understand the origins of the chordates.

○ Interactive cladogram showing the major events that gave rise to the main deuterostome groups. A summary of the key characteristics of each of the deuterostome phyla and subphyla of chordates and hemichordates are combined with an interactive glossary of terms.

Defining Differences

To understand the origins of the phylum Chordata we look at Urochordata and Cephalochordata and how they function. When examining the specimens in the lab or in *Digital Zoology*, these are features that define the phylum and are important for understanding how animals in the phylum function. You'll want to watch for examples of these in the various specimens that you will be looking at in your lab or in *Digital Zoology*.

○ A notochord,
○ postanal segmentation,
○ dorsal hollow nerve cord, and
○ pharyngeal gill slits.

Notochord

The **notochord** gives animals in the phylum Chordata their name and is another example of an innovation in body plans that gave rise to a new group of animals. When you look at a notochord in your lab or in *Digital Zoology*, it seems to be such a simple structure to have had such an important role in the origins in the phylum – a fluid-filled rod of tightly packed cells surrounded by a sheath of elastic connective tissue. You might be more surprised to find that the characteristic that defines the phylum is only present in the most primitive subphyla: urochordates and cephalochordates. In vertebrates the notochord's role as a skeletal structure has been replaced by the vertebral column.

To understand the significance of the notochord you also need to know about the muscles associated with it and how the combination of the two resulted in this new way of moving. Chordates' muscles are arranged segmentally, as **myomeres,** along the length of the notochord with each myomere connected to its neighbor by shared **myosepta**. Muscle fibers in the myomere are aligned in only one direction, anterior to posterior, and there's no fluid-filled coelomic space to act as a **hydrostatic skeleton**. The result is that in the absence of a notochord, when muscle in a myomere contract, it shortens the length of the myomere and stretches the adjacent myomere connected to it by the myosepta. This creates accordionlike ripples down the side of the animal.

When we add the notochord the contraction of the myomeres bends the notochord. When the muscles relax, the elasticity of the notochord returns it to its original straight position, stretching the myomere back to its original length. Repeat the process with the myomeres on the opposite side of the animal and the notochord bends in the opposite direction. When the myomeres relax, the notochord straightens again. By coordinating the contraction of the myomeres down the length of the notocord, the result is the swimming motion typical of the aquatic animals in the phylum with their trunk and tail undulating from side to side, propelling the animal forward. This is a new way of moving.

Postanal segmentation

There's a little more to this than a combination of a notochord and myomeres. The notochord and myomeres extend beyond the anal opening and add more length to this now propulsive structure. This additional length in the body is referred to as **postanal segmentation,** which may sound like a rather technical term for what we're talking about, the first animals with tails. In the other animals that we've seen, the anal opening was always found on the last segment, or most posterior part of the body; this is not so in chordates.

Dorsal hollow nerve cord

Another of the unique features of the chordates is their dorsal hollow nerve cord, which forms from the two sides of the neural plate that fuse together to become the nerve cord lying on top of the notochord and innervating the myomeres that run along its length.

Pharyngeal gill slits

Chordates share their ancestral feeding habit, suspension feeding, with many of the first animals that appeared around the same time. So how do they do it differently? As you might expect, cilia are involved, and they propel the particle-laden water into the digestive tract. However, it's in the most anterior part of the tract, the **pharynx**, where the difference occurs.

In chordates the pharynx is perforated with openings so that water can pass from inside the pharynx through the pharyngeal openings, the slits, to the outside of the animal. The cilia on the walls of the pharynx are moving the water and at the same time trapping food particles contained in the water. Particulate food isn't only trapped by the cilia on the pharynx. An **endostyle** on the ventral surface of the pharynx will also catch food in its mucus net.

A Closer Look Inside *Digital Zoology* - Urochoradata and Cephalochordata

Urochordata: The tunicates

As mentioned tunicates have a **dimorphic life cycle** with the reproductive adult being sessile and the larval stage being free swimming. The urochordates get their common name, tunicates, from the covering, **tunic**, of the sessile adult and the name of the phylum from the notocord found only in the tail of the urochordate larva. Other than the tunic, the other obvious features of the adult are the incurrent and excurrent siphons. Tunicates are suspension feeders, and inside the incurrent siphon is an enlarged perforated **pharynx** that forms a **branchial basket**. This is a ciliated structure, and the cilia on the surface of the basket pull the water into the pharynx and across the **pharyngeal slits** where it is trapped in mucus produced by the endostyle. Like the acorn worm these are **mucociliary** feeders.

To see the remainder of the vertebrate characteristics we need to take a look at the larval stage of the tunicate. Think of it as a smaller version of the branchial basket that we saw in the adult, but this time it's attached to that new way of swimming that appeared in chordates. The tail of the urochordate larva has the notocord, a **dorsal hollow nerve cord,** and small blocks of myomeric muscles that create the undulating swimming motion of the chordates. All of this is located behind the opening of the excurrent siphon of the small branchial basket where the anus is located. It qualifies as **postanal segmentation**, the chordate tail.

Cephalochordata - Branchiostoma: *The lancelet or amphioxus*

The lancelet, *Branchiostoma*, or amphioxus as it is more often called, is a good specimen for seeing many of the chordate characteristics. Whole-mount slides and cross sections are the best way to examine this small fishlike animal. If both aren't available in your lab, they are in *Digital Zoology*.

Segmentation of the body is best seen in the arrangement of the myomeres. In amphioxus each of the myomeres is V-, or chevron-, shaped, and overlaps with the ones on either side. The tail isn't very big in these animals but there are segments after the anal opening, and that's all we need to see to identify this chordate character. As you're looking at the tail be sure that you don't confuse the position of the anal opening with the **atriopore** where water that has crossed the pharyngeal openings leaves the animal.

Amphioxus is also a **mucociliary feeder**. Cilia on the **gill bars** between the gill slits and on the **wheel organ** help to pump water and the particulate food in it into the mouth. Once the water current enters the **pharynx,** mucus secreted by the endostyle traps the particulate materials, and the food-laden mucus is propelled by pharyngeal cilia up and toward the roof of the mouth and from there to the digestive tract. Once the particulate food has been removed, the water passes between the gill bars; through the pharyngeal gill slits; into a cavity, the atrium; and out the atriopore. You'll be able to see some of these details in the whole mounts but cross sections are better. The endostyle does more than create a mucus net; its cells are capable of binding iodine, and it is often thought to be the precursor of the thyroid gland.

The notochord and hollow nerve cord, two other chordate traits, are easy to see in the cross sections and although the notochord is visible in the whole mounts, the nerve cord is a little harder to see. The best way to locate the nerve cord is to use the black photoreceptor cells as landmarks. They are inside the nerve cord. Cephalochordates get their name from the extension of the notocord into the head region.

Blood flow in amphioxus follows the general vertebrate pattern with blood flowing toward the anterior end of the animal inside the ventral aorta. Blood flows through the gill arches to a dorsal aorta and back through the body. As it flows through the gill arches there are multiple blood vessels in each arch, rather than the single blood vessel found in most vertebrates. Unlike vertebrates, there is no heart; instead, the major blood vessels are contractile and used to propel the blood through the system.

The gill bars are much more elaborate in ammocetes. This is related to their increased importance in gas exchange. Ammocetes have gills with lamellae that increase the surface area available for gas exchange. They also have a two-chambered, ventral heart that pumps blood from the ventral aorta over the gills to the dorsal aorta and out to the body.

Cross-References to Other McGraw-Hill Zoology Titles

Integrated Principles of Zoology, 11th edition. C. P. Hickman, L. S. Roberts, and A. Larson. Chapter 25.

Animal Diversity, 3rd edition. C. P. Hickman, L. S. Roberts, and A. Larson. Chapter 15.

Zoology, 5th edition. S. A. Miller and J. P. Harley. Chapter 17.

Biology of the Invertebrates, 4th edition. J. Pechenik. Chapter 22.

Laboratroy Studies in Integrated Principles of Zoology, 10th edition. C. P. Hickman, F. Hickman, and L. Kats. Chapter 17.

General Zoology Laboratory Guide, 13th edition. C. Lytle. Chapter 15.

Structures Checklist

Here are some of the structures that you should be able to easily find in *Digital Zoology* and the specimens that you will be looking at in your lab. After reading your lab handout, you might want to add more and, depending on the equipment available in your lab, you might see more. As you study the material, you might also want to make some notes on how some of these structures looked or include a drawing in your lab notes. (Structures indicated by * may be hard to see.)

Tunicates

Whole mount
- ☐ Adhesive papilla
- ☐ Dorsal fin
- ☐ Endostyle
- ☐ Eye spot
- ☐ Head
- ☐ Nerve cord
- ☐ Notochord
- ☐ Oral siphon
- ☐ Pharynx

- ☐ Statocyst
- ☐ Tail
- ☐ Ventral fin
- ☐
- ☐

Adult
- ☐ Anus
- ☐ Endostyle
- ☐ Excurrent siphon (atriopore)

- ☐ Incurrent siphon (mouth)
- ☐ Intestine
- ☐ Pharynx
- ☐ Stomach
- ☐ Tunic
- ☐
- ☐
- ☐
- ☐
- ☐

- ☐

Additional structures
- ☐
- ☐
- ☐
- ☐

Amphioxis

Whole mount

- ☐ Anus
- ☐ Atriopore
- ☐ Caudal fin
- ☐ Dorsal fin
- ☐ Fin rays
- ☐ Gill bars
- ☐ Gill slits
- ☐ Gonad
- ☐ Hepatic caecum (midgut caecum)
- ☐ Intestine
- ☐ Mouth
- ☐ Myomeres
- ☐ Nerve chord
- ☐ Notochord

- ☐ Oral cirri
- ☐ Oral hood
- ☐ Pigment spots
- ☐ Rostrum
- ☐ Tail
- ☐ Velum
- ☐ Ventral fin
- ☐ Vestibule
- ☐
- ☐
- ☐
- ☐

Cross section

- ☐ Atrium
- ☐ Dorsal aorta
- ☐ Dorsal fin

- ☐ Endostyle
- ☐ Epibranchial groove
- ☐ Fin ray
- ☐ Gill bar
- ☐ Gill slit
- ☐ Gonad
- ☐ Intestine
- ☐ Metapleural folds
- ☐ Myomeres
- ☐ Myosepta
- ☐ Nerve cord
- ☐ Notochord
- ☐ Pharynx
- ☐ Ventral aorta
- ☐ Ventral fin
- ☐

- ☐
- ☐
- ☐
- ☐
- ☐
- ☐
- ☐

Additional structures

- ☐
- ☐
- ☐
- ☐
- ☐

Crossword Puzzle – Urochordata and Cephalochordata

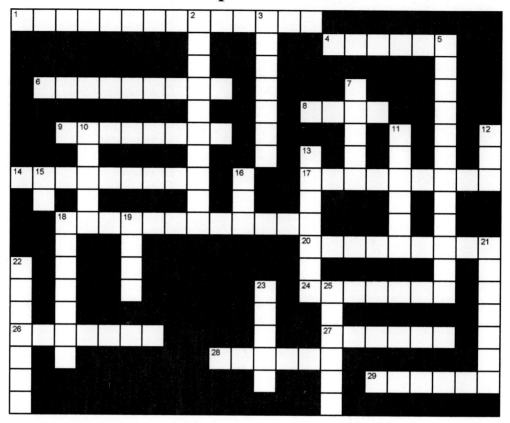

Puzzle solution. An interactive web-based version of this puzzle and its solution are available on the *Digital Zoology* website www.digitalzoology.com/students. With the interactive puzzle you can check to see if individual words or the whole puzzle is correct and get hints for single letters.

Across

1   When characteristics of the immature larval stage ultimately become those of the adult. (14)

4   Myomeres are blocks of this. (6)

6   Cartilaginous supporting rod of a chordate. (9)

8   Maximum number of chambers in a chordate heart. (4)

9   Proposed the most widely accepted version of how the vertebrates arose. (8)

14   This structure on the ventral surface of the urochordate pharynx uses cilia to propel food into the digestive tract. (9)

17   The anus of a urochordate is located next to this siphon. (9)

18   A lancelet pumps its blood through the circulatory system using this structure. (7,5)

20   Urochordates have this common name. (9)

24   Used to describe the V-shaped muscle blocks in a lancelet. (7)

26   Slits in this are one of the chordate characteristics. (7)

27   You'll find more of the ancestral chordate characters in this stage of a urochordate's life cycle. (6)

28   The notochord is found on this side of a chordate. (6)

29   Nonliving covering that surrounds an adult urochordate is secreted by this. (6)

Down

2   Chordates' slits that help separate food from ingested water prior to passing the food into the digestive system. (10)

3   Urochordates have these; one is incurrent, the other excurrent. (7)

5   Unlike arthropods, but similar to echinoderms, chordates have this type of skeleton. (12)

7   Outer covering of an adult urochordate. (5)

10 Position of the nerve cord relative to the notochord in a lancelet. (5)

11 Branchial arteries are also referred to as this type of arch. (6)

12 U-shaped structure in a urochordate. (3)

13 Phagocytic digestion of ingested food occurs in this cephalochordate cecum. (7)

15 Do all chordates have vertebrae? (2)

16 Minimum number of chambers in a chordate heart. (3)

18 Position of the heart in chordates. (7)

19 Another way of describing postanal segmentation. (4)

21 An adult tunicate has this type of existence. (7)

22 The larval stage of a tunicate looks like one of these. (7)

23 Lancelets don't have this in their circulatory system. (5)

25 Describes the structure of the chordate nerve cord. (6)

Self-Test – Urochordata and Cephalochordata

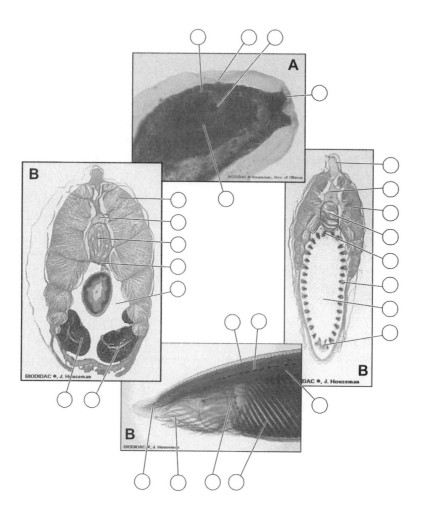

Use the following labels to identify the photographs. You may have to use a label more than once, and some labels may not be appropriate for the photographs. Answers are available on the *Digital Zoology* website at www.digitalzoology.com/students. Be sure to try the interactive drag-and-drop quizzes on the *Digital Zoology* CD-ROM.

Specimens

A- Larval tunicate
B- *Branchiostoma* -
    Amphioxus

Labels

1) Adhesive papillae
2) Atrium
3) Buccal cirri
4) Dorsal fin ray
5) Endostyle
6) Epibranchial groove

7) Gill bar
8) Heart
9) Intestine
10) Metapleural fold
11) Mouth
12) Myomere
13) Nerve cord
14) Notochord
15) Ocelli

16) Oral hood
17) Peripharyngeal band (velum)
18) Pharynx
19) Statocyst
20) Testis

AGNATHA

Inside *Digital Zoology*

As you explore the Agnatha on the *Digital Zoology* CD-ROM or on the web at www.digitalzoology.com, don't miss these learning tools:

○ Photos of whole mounts and cross sections of the ammocoete larva and dissected adult lampreys.
○ Drag-and-drop quizzes on the key structures of the ammocoete larva and the adult lamprey.
○ Interactive cladogram showing the major events that gave rise to vertebrate classes. A summary of the key characteristics of each class is combined with an interactive glossary of terms.

Defining Differences

Agnathans, which include the lampreys and hagfish, are often described by what they don't have: vertebrae, paired lateral fins, and jaws. It's the absence of these characteristics that is often used as proof that they are neither vertebrates nor fish. On the other side of the argument is the presence of a cranium protecting the brain and a vertebrate trait; they do swim, which makes them a fish. Some of the differences described in the following sections appear for the first time in the Agnatha, and they define the superclass of vertebrates. Others are important for understanding how agnathans function. Whichever is the case, you'll want to watch for examples of these in the specimens that you will be examining.

○ A cyclostome, a
○ cartilaginous skeleton and persistent notochord,
○ cranium, and a
○ pineal eye.

Cyclostome

Unlike the rest of the vertebrates, an agnathan doesn't have jaws. This is how they get their name, which means jawless. Without a jaw an agnathan's mouth is always open and circular, the origin of a second term, Cyclostomata, a taxonomic term sometimes used to combine the two living groups of agnathans-- lampreys and hagfish. Not everyone agrees on how closely the two are related; you'll find Agnatha used, as a valid taxon, usually a superclass, or as a common name to describe the two groups. Hagfish are primarily scavengers feeding on invertebrates buried in the mud or on dead or dying fish. The absence of a jaw makes it difficult to bite off pieces of food. Therefore, hagfish burrow into their food through the mouth, anus, or gills and eat their meal from the inside out. The tongue is covered with small teeth that combine with a cartilaginous plate to rip off small pieces of flesh that can be swallowed. Lampreys are ectoparasites of other fish, and their tongue and circular oral disk are covered in teeth that help the lamprey hold onto its prey while it feeds on blood and other fluids that pool around the wound. Both animals are faced with a problem of breathing with a mouth full of food, and the hagfish's single nostril allows water to flow into the pharynx and across the gill sacs. When attached to their host, lampreys pump water in and out of the gill sacs rather than having it flow through the mouth and pharynx; through the gill sac; and out the external opening, the usual path for water when not attached to the host.

Cartilaginous skeleton and persistent notochord

Both hagfish and lampreys bear little or no resemblance to their ancestors who were heavily armored fish covered with bony plates instead of scales. Fish that had a jaw to hold onto their prey and who were protected with individual bony scales that were more flexible than plates replaced the ancient agnathans. Unlike ancestral agnathans, hagfish and lampreys have a glandular skin without scales and use a diffuse meshwork of cartilage to support the openings to the gill sacs. There's no sign of **vertebrae**, although small cartilaginous blocks along the length of the **notochord** in lampreys may be **homologous** vertebrae. In both agnathans, the notochord is still present in the adult, the only vertebrates where this is the case, and it still serves its original function.

✍   *What is the role of the notochord in chordates?*

Cranium

The **cranium** is a protective case that surrounds and protects the brain and the sensory structures located on the head. Typically it is composed of three parts: the **splanchnocranium** includes skeletal elements that support and hold the gill slits open, the **chondrocranium** forming the supporting structures under the brain, and the **dermatocranium** forming the outer casing of the skull. This last part of the cranium is missing in agnathans and in most vertebrates is the bone that replaces the splanchnocranium and chondrocranium.

Pineal eye

Lampreys have a third eye, the **pineal eye**, located on the dorsal midline behind the nostril. The structure appears to be unpaired but is two pineal eyes that have fused together. It's near enough to the surface to detect light, and it may control **circadian rhythms** by secretion of **melatonin**. A similar structure is present in hagfish but below the surface of the body. The pineal retains its secretory function and is more appropriately called the pineal gland because it doesn't develop into an eyelike structure as it does in lampreys.

A Closer Look Inside *Digital Zoology* - Agnatha

Ammocoete larva

The larval stage of the lamprey is called an **ammocoete larva**, and although it looks somewhat like amphioxus don't forget that it is a vertebrate. Look for the chordate characters in the whole-mount slide and cross sections. The hollow nerve cord is located above the notochord, and it's enlarged at the anterior end to form a brain with three lobes, each with separate functions. The forebrain is olfactory and connects with the nostril. The midbrain is visual and connects to a pair of eyes, which in the larva aren't functional. These larvae spend most of their time with their heads buried in the mud, and instead of using a pair of anterior eyes to detect light they have special photoreceptors in the tail that sticks out of the mud. The hindbrain is connected to small ear vesicles and the remainder of the nerve cord. The **myomeres** in this animal are much smaller than those of amphioxus and don't wrap around the sides of the body in the same way. You may find them easier to see in the sections rather than the small whole mount.

Ammocoetes also have **pharyngeal gill slits** but their shape and number have changed from what we saw in amphioxus. They're no longer U-shaped, and there are only seven pairs. There is one other fundamental difference. **Cilia** are no longer involved in pumping water through the **pharynx**. Instead, pharyngeal musculature and the **velum** change the shape of the buccal cavity to suck in food. Ammocoetes are still mucociliary feeders, and as water is forced through the gill slits food is trapped in mucus produced by goblet cells on the gill bars. The **endostyle** still secretes mucus that helps to trap food but it is now enclosed in a tube that runs along the floor of the pharynx, and the mucus it produces contains iodinated compounds. During **metamorphosis** it becomes the thyroid gland of the adult lamprey. Cilia are still important in feeding, and they move food into the digestive tract as they did in the early chordates.

The gill bars are elaborate in ammocoetes. This is related to their increased importance in gas exchange. Ammocoetes have gills with lamellae that increase the surface area available for gas exchange. They also have a two-chambered, ventral heart that pumps blood from the ventral aorta over the gills to the dorsal aorta and out to the body

Petromyzon: *The lamprey*

Although the agnathans are an ancient lineage with a larval stage that shows us some of the early chordate characters, the adults are unique ectoparasites of fish with a life cycle that can take them from freshwater to the oceans and back.

 What are the main stages in the lamprey life cycle?

The body is eel-like and almost circular with medial dorsal and caudal fins and none of the paired lateral fins that we normally associate with fish. That's not the only difference. If you feel the surface you'll notice that it's smooth; lampreys have a glandular skin not protected by the usual scales found on fish. The most obvious external feature is the oral disk and oral funnel. Without a jaw to clamp down on their prey, the oral disk and its teeth create the suction to hold the lamprey in place. The tongue is also covered in keratinized teeth, and its movements grind away at the surface of the host to which the lamprey is attached. Blood and other fluids that flow from the open wound pass through the mouth located at the base of the oral funnel and just above the base of the tongue. Above the head and on the midline are two unpaired structures, the nostril and pineal eye. On the sides of the head are paired eyes and seven pairs of gill slits, the openings to the gill pouches.

The lamprey is usually dissected along the midline with a midsagittal cut and occasionally by cross sections at different points down the length of the body. You'll find both in *Digital Zoology*. As you make the sagittal cut through the animal you'll notice that although scales are missing, the body is still rigid and tough due to the cartilaginous skeleton. The internal skeleton forms a supporting net for the branchial apertures and cranium. A **lingular cartilage** supports the tongue, and you may be able to see traces of the cartilages that support the gill slits. Lampreys, unlike any of the other chordates, retain the notochord in the adult, and it runs the length of the body, above the gut and below the dorsal nerve cord. Some landmarks in the section to watch for are the **olfactory sac** that opens to the **nostril**, the two-chambered heart, the pineal eye, and the **liver**. The digestive system extends from the mouth to the pharynx and directly to the intestine that passes between the liver and the unpaired gonad before becoming a straight tube that opens in the cloaca. Lampreys, like most fish, don't have a stomach.

If you don't have cross sections available in your lab use the ones in *Digital Zoology* to compare to your sagittal section.

Cross-References to Other McGraw-Hill Zoology Titles

Integrated Principles of Zoology, 11th edition. C. P. Hickman, L. S. Roberts, and A. Larson. Chapter 26.

Animal Diversity, 3rd edition. C. P. Hickman, L. S. Roberts, and A. Larson. Chapter 16.

Zoology, 5th edition. S. A. Miller and J. P. Harley. Chapter 18.

Laboratory Studies in Integrated Principles of Zoology, 10th edition. C. P. Hickman, F. Hickman, and L. Kats. Chapter 18.

Vertebrate Biology, 1st edition. D.W. Linzey. Chapter 4.

Structures Checklist

Here are some of the structures that you should be able to easily find in *Digital Zoology* and the specimens that you will be looking at in your lab. After reading your lab handout, you might want to add more and, depending on the equipment available in your lab, you might see more. As you study the material, you might also want to make some notes on how some of these structures looked or include a drawing in your lab notes. (Structures indicated by * may be hard to see.)

Larval lamprey - Ammocoete larva

Whole mount

- ☐ Anus
- ☐ Caudal fin
- ☐ Dorsal fin
- ☐ Endostyle (Subpharyngeal gland)
- ☐ Esophagus
- ☐ Forebrain
- ☐ Gill bars
- ☐ Gill pouch
- ☐ Gill slits
- ☐ Hindbrain
- ☐ Intestine
- ☐ Midbrain
- ☐ Nerve cord
- ☐ Notochord
- ☐ Oral hood
- ☐ Oral papillae
- ☐ Pronephric ducts
- ☐ Tail
- ☐
- ☐
- ☐
- ☐

Cross section

- ☐ Dorsal aorta
- ☐ Ear vesicle
- ☐ Endostyle (subpharyngeal gland)
- ☐ Gill bars
- ☐ Gill lamellae
- ☐ Gill slit
- ☐ Hyperpharyngeal ridge
- ☐ Hypopharyngeal ridge
- ☐ Kidney
- ☐ Intestine
- ☐ Myomeres
- ☐ Myosepta
- ☐ Nerve cord
- ☐ Notochord
- ☐ Pharynx
- ☐ Typhlosole
- ☐
- ☐
- ☐
- ☐
- ☐

Additional structures

- ☐
- ☐
- ☐
- ☐
- ☐

Adult lamprey

External anatomy

- ☐ Anterior dorsal fin
- ☐ Anus
- ☐ Buccal funnel
- ☐ Buccal papillae
- ☐ Caudal fin
- ☐ Eye
- ☐ Gill slits
- ☐ Lateral line
- ☐ Nostril
- ☐ Pineal organ
- ☐ Posterior dorsal fin
- ☐ Teeth
- ☐ Tongue
- ☐
- ☐
- ☐
- ☐

Cross section

- ☐ Annular cartilage
- ☐ Anterior cardinal vein
- ☐ Brachial sinus
- ☐ Branchial artery
- ☐ Branchial tube
- ☐ Carotid artery
- ☐ Caudal artery
- ☐ Caudal vein

- ☐ Esophagus
- ☐ Eye
- ☐ Fin ray cartilages
- ☐ Gill pouches
- ☐ Gills
- ☐ Gonad
- ☐ Inferior jugular vein
- ☐ Intestine
- ☐ Kidney
- ☐ Lens
- ☐ Lingular cartilage
- ☐ Lingular muscle
- ☐ Liver
- ☐ Myomeres
- ☐ Nasopharyngeal pouch
- ☐ Notochord
- ☐ Ocular muscles
- ☐ Olfactory sac
- ☐ Pharyngeal gland
- ☐ Pharyngeal muscle
- ☐ Pharynx
- ☐ Pineal organ
- ☐ Posterior cardinal vein
- ☐ Retina
- ☐ Spinal cord (nerve cord)
- ☐ Ventral aorta

- ☐
- ☐
- ☐
- ☐

Sagittal section

- ☐ Annular cartilage
- ☐ Brain
- ☐ Branchial tube
- ☐ Buccal funnel
- ☐ Buccal muscle
- ☐ Cranial cartilage
- ☐ Dorsal aorta*
- ☐ Esophagus
- ☐ Gill pouch
- ☐ Hard palate
- ☐ Heart
- ☐ Inferior jugular vein*
- ☐ Intestine
- ☐ Lingual cartilage
- ☐ Liver
- ☐ Mouth
- ☐ Myomeres
- ☐ Nasopharyngeal pouch
- ☐ Nostril
- ☐ Notochord
- ☐ Pharynx

- ☐ Pineal organ
- ☐ Posterior cardinal vein
- ☐ Protractor muscle of the tongue
- ☐ Sinosus venosus
- ☐ Spinal cord (nerve cord)
- ☐ Spiral valve
- ☐ Tongue
- ☐ Velum
- ☐ Ventral aorta*
- ☐ Ventricle
- ☐
- ☐
- ☐

Additional structures

- ☐
- ☐
- ☐
- ☐
- ☐
- ☐
- ☐

Crossword Puzzle – Agnatha

Puzzle solution. An interactive web-based version of this puzzle and its solution are available on the *Digital Zoology* website www.digitalzoology.com/students. With the interactive puzzle you can check to see if individual words or the whole puzzle is correct and get hints for single letters.

Across

1 Another term describing that anterior swelling of the dorsal nerve cord. (5)

4 One type of agnathan is the lamprey, the other is this. (7)

8 Name given to the blocks of muscle along the length of the lamprey or hagfish. (8)

9 Hagfish are specialists at producing this protective substance. (5)

12 This part of most vertebrate digestive systems is missing in agnathans. (7)

13 The main skeletal support in an agnathan. (9)

15 Number of semicircular canals in a hagfish. (3)

16 Slows the passage of food through the digestive tract of a lamprey. (6,5)

19 Number of chambers in an agnathan's heart. (3)

20 Paired structures found on other fish, but not on the agnathans. (4)

23 The change from a larval to an adult. (13)

26 An agnathan's skeleton is composed of this. (9)

27 Hagfish feed on dead and dying fish; they are best described as this instead of being parasites. (10)

29 This type of pesticide controls the lamprey in the Great Lakes watershed. (10)

31 We refer to the body of an agnathan as _____ like. (3)

32 Found on the tongue of a hagfish and help it consume its meal. (5)

33 The name for a larval lamprey. (9)

34 Food ingested by the larval ammocoete is moved into the digestive tract by cilia on this structure in the floor of the mouth. (9)

35 Were there any lampreys in the Great Lakes in the nineteenth century? (2)

Down

1 You won't find these skeletal elements in an agnathan. (5)

2 After passing across the ammocoete gills, water moves to the atrium and out this structure. (9)

3 The respiratory surface of a lamprey is located inside these. (7)

5 Larval lampreys are found in this type of environment. (10)

6 External opening of the agnathan respiratory system. (7)

7 Unlike other fish, you won't find these on the surface of an agnathan. (6)

10 The slime from a hagfish has two possible effects on fish that attack them. One is that it suffocates the attacker; this is the other mode of action. (8)

11 Unlike the other vertebrates, hagfish blood and seawater are described as this. (8)

14 Hagfish clean themselves by tying themselves into this. (4)

17 Formal name for jawless fish. (7)

18 Parasite that lives on the surface of its host is what type of parasite? (12)

20 Lampreys are best described as this type of feeder. (5)

21 Although they don't have vertebrae, agnathans do have a skeletal structure called this and are included in the taxon because of it. (7)

22 When a lamprey feeds, water enters and exits through these to ventilate the gill pouches. (4,5)

24 Paired sensory structures in lampreys missing in hagfish. (4)

25 In addition to photoreceptive eyes, lampreys also have this optic device. (6)

28 Cilia on this organ create the water currents that bring food into the ammocoete's mouth. (5)

30 Number of external gill slits in a lamprey. (5)

Self-Test – Agnatha

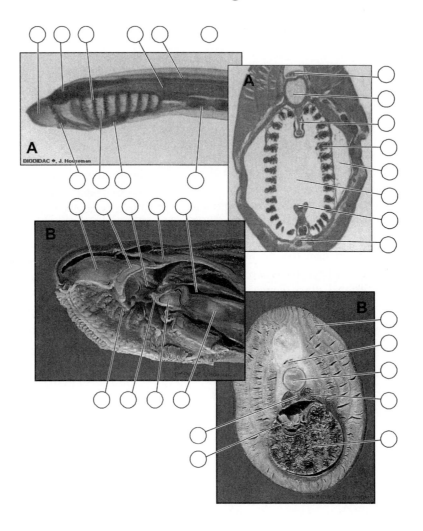

Use the following labels to identify the photographs. You may have to use a label more than once, and some labels may not be appropriate for the photographs. Answers are available on the *Digital Zoology* website at www.digitalzoology.com/students. Be sure to try the interactive drag-and-drop quizzes on the *Digital Zoology* CD-ROM.

Specimens

A- Laraval lamprey
(ammocoete larva)

B- Adult lamprey

Labels

1) Annular cartilage
2) Brain
3) Buccal muscle
4) Cranial cartilage
5) Dorsal aorta
6) Endostyle
7) Gill
8) Gill chamber
9) Gill slit
10) Heart

11) Hyperpharyngeal ridge
12) Hypopharyngeal ridge
13) Intestine
14) Lingular cartilage
15) Liver
16) Mouth
17) Myomere
18) Nerve cord
19) Notochord
20) Oral hood

21) Oral papilla
22) Pharynx
23) Posterior cardinal vein
24) Tongue
25) Tooth

JAWED FISHES

Inside *Digital Zoology*

As you explore the jawed fishes on the *Digital Zoology* CD-ROM or on the web at www.digitalzoology.com, don't miss these learning tools:

○ Photos of dissected specimens of cartilaginous and bony fishes, which includes the shark and perch, respectively.

○ Drag-and-drop quizzes on the external anatomy and skeleton, muscular systems, and internal anatomy for the shark and perch.

○ Interactive cladogram showing the major events that gave rise to vertebrate classes. A summary of the key characteristics of each class are combined with an interactive glossary of terms.

Defining Differences

Fish is a pretty broad term that includes anything that's vertebrate, cold-blooded, swims, and has fins. In this part of *Digital Zoology* we take a look at two different types of fish: cartilaginous fish, Chondrichthyes, and bony fish, Osteichthyes. Each of the features described in the following list applies to the two classes and, depending how the feature is modified, it may define differences between the two classes of jawed fishes. Whichever is the case, you'll want to watch for examples of these in the specimens you will be examining.

○ Jaws,
○ gills,
○ paired and unpaired fins with a fusiform body covered in
○ scales.

Jaws

We see **jaws** for the first time in living vertebrates with the cartilaginous fish, Chondrichthyes. These, and all other "jawed-mouthed" vertebrates, are collectively referred to as the **gnathostomes**. Their counterparts are vertebrates with a mouth, but no jaws – the **agnathans,** hagfish and lampreys. Both groups descended from early fishes that were nothing like what we see today. **Ostracoderms,** the agnathan relative, and **placoderms,** ancestor to modern-day jawed fishes, were heavily armored with large **dermal skeletal plates** covering their bodies. Although they dominated the oceans at one time, these ancient fish started to disappear as earth became much warmer and drier at the end of the Devonian. (In the next chapter we'll see how these same conditions started the vertebrate transition to land and the origins of the Amphibia.) Whatever the reason is for the disappearance of the armored fish they were replaced by cartilaginous and bony fish that resemble those we see today. As you look at each of the two groups during your lab, or in *Digital Zoology*, you'll see hints about their evolutionary past. Remember the two groups diverged long ago; each is highly specialized. Don't look at the modern-day Chondrichthyes as an ancestor to the bony fishes; it didn't happen that way.

Jaws were a major innovation for the first animals that had them. They could feed more aggressively by sucking up large amounts of substrate, trapping and holding prey in place, or taking a bite out of their prey. As we've seen, any innovation that allows for a new way of feeding usually results in diversification of the group. That's what happens with the gnathostomes. So how do biologists believe that jaws first appeared?

As we've seen in the previous chapter, early chordates fed on particulate food pulled into the mouth by **cilia** inside the pharynx. The same cilia filtered the particulate food from the water sending food into the digestive system and water out through **gill slits** in the side of the **pharynx**. The openings in the wall of the pharynx were held open by **pharyngeal arches**.

There's another way to move particle-laden water through the pharynx that doesn't require cilia. The pharyngeal cavity becomes the pump, shrinking to force water out through the gill slits on the sides, then increasing in size to pull water in through the mouth. The elasticity of the pharyngeal arches plays a key role in the mechanism. When the pharyngeal muscles contract, the arches bend, and when the muscles relax, the arches spring back to their original shape. This returns the pharyngeal cavity to its original size and pulls in water. Many of the first fishes, ostracoderms for example, used this pharyngeal pumping strategy to feed. They placed their mouths against the substrate and sucked it in and became substrate, rather than suspension, feeders.

There's still some room for improvement. If the diameter of the mouth remains the same as the pharyngeal cavity pumps, as much water is likely to go out of the mouth as come in the mouth. If the diameter of the mouth is decreased at the same time the pharyngeal cavity decreases in size, then more water would be forced out through the pharyngeal slits than through the mouth. The most anterior pharyngeal arch, already capable of bending, bent more than the rest of the arches and the mouth started to close. We have the beginnings of the rudimentary gnathostome jaw as a modification of the first pharyngeal (gill) arch.

Gills

Pharyngeal arches, like all tissues, are supplied with blood vessels. With oxygenated water flowing past them, the arches became an ideal place to extract oxygen from water. The result was gills with a supporting **gill arch**, a **gill raker** on the inside of the arch, and a pair of **gill filaments** on the outside. Once cilia were no longer involved in feeding, the gill rakers kept the ingested food inside the pharynx and let the water flow out and over the gill filaments. Gill filaments are made of minute **lamella** and are the site of gas exchange. Water flows across these lamellae in the opposite direction to the blood inside, in a pattern referred to as **countercurrent exchange**.

 Why is countercurrent flow of blood and water better than the concurrent flow of the two fluids?

Other than swimming with an open mouth and changing the shape of the buccal cavity, the cartilaginous fish had no other way to pump water across their gill surfaces. Bony fish developed a way to do that using an **operculum**, a bony gill covering. The operculum not only protects the gills that lay underneath it, it is hinged so that its posterior edge can be pivoted away from the body. As it moves away from the body it pulls water from the buccal cavity out and across the gill surface. The movement of the operculum is coordinated with the opening and closing of the mouth and changes in the shape of the buccal cavity creating a steady flow of water across the gills.

Paired and unpaired fins with a fusiform body

The aquatic environment is a lot denser than the terrestrial, and animals that live in water face special challenges when they are trying to move. Compare running through a pool of waist-deep water to running along the deck beside the pool and you'll get the idea. The best way to move through a dense medium is to streamline the body, and that's what fish have done. Their bodies are **fusiform** with the tallest and widest part in the middle and the body tapering toward the anterior and posterior ends of the animal.

Unlike today's fish, the earliest fish didn't have any fins, and if they moved slowly, it probably wasn't a problem. As they moved faster, there was a potential for accidentally tilting, twisting, or turning and stability became a problem. The solution was fins, either unpaired or paired, which extended along the dorsal ventral line of the body and from the sides of the body. The unpaired fins include the **dorsal fin, anal fin,** and **caudal fin**. The **pectoral fins** in the front and **pelvic fins** behind them are both paired and attached to skeletal elements, **girdles**, that connect the right and left sides. But more than connecting the two sides, the girdles are an important site for attachment of the muscles that moved the fins.

 What are the different types of caudal fins, and which types are found in the two classes of jawed fish?

Scales

Vertebrate skin has the two layers: an outer **epidermis** and an underlying **dermis**. In fish, the epidermis is living and contains **mucus glands,** and the dermis underneath produces the **scales** that cover the body. Ancestrally the entire surface of the fish was covered in **dermal bone,** but in living fish species, scales have replaced this. Think of it as the change from solid metal armor that protected medieval warriors to the use of chain-link armor with its interconnected metal loops, although in the case of our fish, the scales are solid, not hollow. Some fish don't have scales or they're reduced to only certain parts of the body. In these fish, secretions of the mucus gland in the epidermal layer may help as a defense against predators.

A Closer Look Inside *Digital Zoology* - Jawed Fishes

Chondrichthyes: Dogfish shark

Sharks are cartilaginous fish and, as the name implies, their skeleton is made of **cartilage**. It's not clear if this is an ancestral or derived trait. The earliest fishes, including those identified as the ancestors to the sharks, the placoderms, were covered with bony dermal plates. This suggests that the presence of a purely cartilaginous skeleton is a derived trait. That may not be the whole story, though. The fossil record is biased for bony animals, not cartilaginous ones, and maybe there is a cartilaginous ancestor that has yet to be identified.

If you rub the skin of a shark, it feels like sandpaper, and dried shark skin has been used as sandpaper. Its texture results from **placoid scales**, **denticles**, that cover a shark's body. Like the scales of other fish, placoid scales are formed in the **dermal layer** of the skin, but they differ by piercing through the epidermis with a backwards projecting spine. The scale is made of **enamel**-covered **dentine,** and its large hollow base is embedded deep in the underlying dermis. The scales are important for reducing friction and drag as water flows over the surface of the body. Under ideal circumstances, friction and drag are at their lowest with laminar flow of the water. Surprisingly, perfectly smooth surfaces disrupt laminar flow, and a shark's scales direct the movement of water over the body to help maintain laminar as the shark swims. Competitive swimmers have picked up on this, and Speedo's "fastskin" full body bathing suit, which made its appearance at the Sydney Summer Olympic Games, is designed on the same principles as shark skin. A shark's teeth are larger versions of these scales.

✍ *What is a shark's spiral valve, and what does it do?*

The urogenital system in the shark gives us a chance to see the close relationship between the reproductive and excretory systems and how they were organized in early vertebrates. The two systems are closely linked to each other, sharing the same ducts and tubes, and are embryological neighbors forming from adjacent tissues. Starting at the anterior end of the embryo, the **nephric ridge** differentiates forming tubules that will ultimately be **nephrons**. These are segmentally arranged down the length of the embryo and connected to each other by the **archinephric duct,** which drains to the outside. The entire length has the potential to be the kidney but kidneys form from parts of the ridge. Vertebrate kidneys form from either that anterior region, **pronephros**; middle region, **mesonephros**; or the posterior region, **metanephros**. The sharks urogenital system is an example of **opisthonephros,** a modified mesonephros extended posteriorly by the addition of more tubules.

✍ *What is the main metabolic waste in sharks, and how is it used for osmoregulation?*

Ancestrally, gonads developed from the tissues beside the kidney and shared the archinephric duct with the kidney to transport gametes outside the animal. This is still the case in the male shark, and the **testes** use the tubules at the anterior end of the kidney, no longer involved in excretion. Only the more posterior end of the male kidney is excretory, and it dumps its wastes into a **urinary duct** leaving the original archinephric duct to be used by the reproductive system. The close association between the excretory and reproductive systems that occurs in males doesn't occur in females. Females have a separate tube, the **oviduct**, to transport their **eggs**. **Ovaries** shed their eggs into the coelom, and they are picked up by the ciliated funnel at the anterior end of the oviduct, the **oviducal funnel**. The eggs move down the oviduct by a combination of muscular contractions and cilia. The oviduct may have specialized regions for adding shell, **shell gland**, or other membranes to the egg. In the some sharks, such as the dogfish shark, the posterior end of the oviduct is modified into an **ovisac** where the egg capsule can develop internally.

Osteichthyes: Perch

Bony fish have two different types of scales, **cycloid** and **ctenoid**. The two are similar in composition and differ only in their appearance. Cycloid scales are rounded with smooth edges; ctenoid scales have small projections or teeth that stick out from the epidermal covering. These small teeth may work the same way that the placoid scales do in sharks, enhancing the laminar flow of water over the body as the fish swims. There are concentric rings in both scales, and they are used to determine a fish's age.

Because living tissue is denser than water a fish would sink to the bottom if it didn't keep swimming or find some way to achieve the same buoyant density as the water that surrounds the fish. The bony fish achieve this with an air-filled **swim bladder**. It's not enough to have an internal bag of air. The amount of air in the bag has to change. As a fish swims deeper, the weight of the column of water above increases and compresses the air inside the swim bladder. More air will need to be pumped into the bladder to achieve neutral buoyancy at greater depths. It's the opposite as the fish rises to the surface; air will have to be removed from the bladder. Swim bladders evolved from the lungs of the early lungfish, and in today's bony fish, they are either connected to the alimentary tract by a pneumatic duct, **phystostomas,** or are disconnected, **physoclistous**. Phystostomas fish let air out through the pneumatic duct. Physoclistous fish can't do that. They have a highly vascularized surface of the bladder, the **ovale**, where gas moves from the bladder to the blood. Both types of fish add air to the bladder using the **gas gland** and its network of capillaries, the **rete mirable**, that secrete oxygen from the blood to the bladder.

✍ *How does a shark achieve a neutral buoyancy?*

Achieving neutral buoyancy was important and had a tremendous impact on the evolution of the bony fishes. A shark's **heterocercal** tail was combined with the pectoral and pelvic fins to constantly generate lift. If a shark stopped swimming, it would sink. With a swim bladder this doesn't happen in bony fish. Their tail is **homocercal** and the pectoral and pelvic fins can be used for something other than generating lift as they did in the shark. Bony fishes' fins are supported by dermal bone that forms **fin rays,** and musculature associated with the rays allows the fins to fold and unfold. This is best seen in the pectoral fin, which in bony fish is attached to the head. This fin bends, folds, and moves in a variety of ways that allow more precise and delicate movements associated with hovering.

The urogenital system of the bony fish has been modified from what we saw in the shark. Although they have an **opisthonephric kidney,** the excretory and reproductive systems are separate. In females, the **oviduct** and its **oviducal funnel** have been replaced with an **ovarian duct** that connects the ovary directly to the genital opening. Males have also developed a separate **testicular duct,** and the **archinephric duct** is used exclusively by the kidney.

Cross-References to Other McGraw-Hill Zoology Titles

Integrated Principles of Zoology, 11th edition. C. P. Hickman, L. S. Roberts, and A. Larson. Chapter 26.

Animal Diversity, 3rd edition. C. P. Hickman, L. S. Roberts, and A. Larson. Chapter 16.

Zoology, 5th edition, S. A. Miller and J. P. Harley. Chapter 18.

Vertebrate Biology, 1st edition. D. W. Linzey. Chapter 5.

Laboratory Studies in Integrated Principles of Zoology, 10th edition. C. P. Hickman, F. Hickman, and L. Kats. Chapter 18.

General Zoology Laboratory Guide, 13th edition. C. Lytle. Chapters 16 & 17.

Structures Checklist

Here are some of the structures that you should be able to easily find in *Digital Zoology* and the specimens that you will be looking at in your lab. After reading your lab handout, you might want to add more and, depending on the equipment available in your lab, you might see more. As you study the material, you might also want to make some notes on how some of these structures looked or include a drawing in your lab notes.

Dogfish shark - External anatomy and skeletomuscular systems

External anatomy

- [ ] Abdominal pores
- [ ] Ampullae of Lorenzini
- [ ] Caudal fin
- [ ] Clasper (male)
- [ ] Cloaca
- [ ] Dermis
- [ ] Dorsal fins (anterior and posterior)
- [ ] Endolymphatic pores
- [ ] Epidermis
- [ ] Eyes
- [ ] Fin rays
- [ ] Fin spines
- [ ] Gill slits
- [ ] Head
- [ ] Labial pouch and groove
- [ ] Lateral line
- [ ] Mouth
- [ ] Naris (nostril)
- [ ] Pectoral fin
- [ ] Pelvic fin
- [ ] Placoid scales (denticles)
- [ ] Rostrum
- [ ] Spiracle
- [ ] Trunk
- [ ] Urinary papilla (female)
- [ ] Urogenital papilla (male)
- [ ]
- [ ]
- [ ]

Muscles

- [ ] First levator (palatoquadrate)
- [ ] Common coracoarcuals
- [ ] Coracobranchials
- [ ] Coracohyoid
- [ ] Coracomandibular
- [ ] Depressor of pectoral fin
- [ ] Dorsal constrictors
- [ ] Epaxial muscles
- [ ] Epibranchial musculature
- [ ] Horizontal septum
- [ ] Hypaxial muscles
- [ ] Interhyoid
- [ ] Levator hyomandibulae
- [ ] Levator of pectoral fin
- [ ] Linea alba
- [ ] Mandibular adductor
- [ ] Myomere
- [ ] Myosepta
- [ ] Radial muscles
- [ ] Spiracular muscle
- [ ] Ventral constrictors
- [ ]
- [ ]
- [ ]
- [ ]

Skeleton - Skull

- [ ] Antorbital process
- [ ] Antorbital shelf
- [ ] Basal plate
- [ ] Basibranchial cartilages
- [ ] Basihyal cartilage
- [ ] Ceratobranchial cartilages
- [ ] Ceratohyal cartilage
- [ ] Endolymphatic foramina
- [ ] Endolymphatic fossa
- [ ] Epibranchial cartilages
- [ ] Foramen magnum
- [ ] Hyomandibular cartilages
- [ ] Hypobranchial cartilages
- [ ] Infraorbital shelf
- [ ] Labial cartilage
- [ ] Meckel's cartilage
- [ ] Occipital condyles
- [ ] Olfactory capsules
- [ ] Palatopterygo-quadrate cartilage
- [ ] Pharyngobranchial cartilages
- [ ] Postorbital process
- [ ] Postotic process
- [ ] Precerebral cavity
- [ ] Quadrate process
- [ ] Rostral fenestrae
- [ ] Rostral carina
- [ ] Rostrum
- [ ] Superficial ophthalmic foramina
- [ ] Supraorbital crest
- [ ] Supraotic crest
- [ ] Trochlear foramen
- [ ] Vagus foramina
- [ ]
- [ ]
- [ ]

Skeleton - Trunk and appendages

- [ ] Acetabular surface
- [ ] Basal cartilages
- [ ] Caudal vertebrae
- [ ] Centrum
- [ ] Ceratotrichia
- [ ] Clasper cartilages male)
- [ ] Coracoid bar
- [ ] Glenoid surface
- [ ] Hemal canal
- [ ] Hemal spine
- [ ] Iliac process
- [ ] Intercalary plate
- [ ] Ischiopubic bar
- [ ] Mesopterygium
- [ ] Metapterygium
- [ ] Neural canal
- [ ] Neural spine
- [ ] Propterygium
- [ ] Radial cartilages
- [ ] Scapular cartilage
- [ ] Suprascapular cartilage
- [ ] Trunk vertebrae
- [ ]
- [ ]

Dogfish shark - Internal anatomy

Digestive and related systems

- [ ] Bile duct
- [ ] Buccal cavity
- [ ] Cardiac region of stomach
- [ ] Cloaca
- [ ] Colon
- [ ] Duodenum
- [ ] Esophageal papillae
- [ ] Gall bladder
- [ ] Gastrosplenic ligament
- [ ] Hepatoduodenal ligament
- [ ] Hepatogastric ligament
- [ ] Ileum
- [ ] Lesser omentum
- [ ] Liver: right, medial, and left lobes
- [ ] Mesentery
- [ ] Mesorectum
- [ ] Pancreas, dorsal lobe
- [ ] Pancreas, ventral lobe
- [ ] Pyloric region of stomach
- [ ] Rectal gland
- [ ] Rectum
- [ ] Rugae of stomach
- [ ] Spiral valve
- [ ] Spleen
- [ ] Teeth
- [ ] Tongue
- [ ]
- [ ]
- [ ]
- [ ]
- [ ]

Circulatory and respiratory systems

- [ ] Annular arteries
- [ ] Annular veins
- [ ] Anterior cardinal vein
- [ ] Atrium
- [ ] Branchial arteries, afferent
- [ ] Branchial arteries, efferent
- [ ] Caudal artery
- [ ] Caudal vein
- [ ] Celiac artery
- [ ] Common cardinal veins
- [ ] Conus arteriosus
- [ ] Coronary artery
- [ ] Demibranchs
- [ ] Dorsal aorta
- [ ] Esophageal artery
- [ ] Femoral vein
- [ ] Gastric artery
- [ ] Gastric vein
- [ ] Gastrohepatic artery
- [ ] Gastrosplenic artery
- [ ] Gill arch
- [ ] Gill filaments
- [ ] Gill rakers
- [ ] Gill rays
- [ ] Hepatic artery
- [ ] Hepatic portal vein
- [ ] Hepatic vein
- [ ] Holobranch
- [ ] Iliac arteries
- [ ] Iliac vein
- [ ] Inferior jugular vein
- [ ] Internal carotid artery
- [ ] Internal gill slits
- [ ] Intestinal arteries (anterior and posterior)
- [ ] Intestinal veins (anterior and Posterior)
- [ ] Lateral abdominal veins
- [ ] Lienomesenteric vein
- [ ] Mesenteric arteries
- [ ] Mesenteric veins
- [ ] Pancreaticomesenteric artery
- [ ] Pancreaticomesenteric vein
- [ ] Pericardial cavity
- [ ] Posterior cardinal vein
- [ ] Renal veins
- [ ] Sinus venosus
- [ ] Subclavian artery
- [ ] Subclavian vein
- [ ] Ventricle
- [ ]
- [ ]
- [ ]
- [ ]
- [ ]

Urogenital system

- [ ] Cloaca
- [ ] Efferent ductules
- [ ] Kidneys
- [ ] Mesonephric ducts
- [ ] Mesorchium
- [ ] Mesotubarium
- [ ] Mesovarium
- [ ] Ostium tubae
- [ ] Ovaries
- [ ] Oviducts
- [ ] Seminal vesicle
- [ ] Shell gland
- [ ] Testes
- [ ] Urinary papilla
- [ ] Urogenital papilla
- [ ]
- [ ]
- [ ]

Additional structures

- [ ]
- [ ]
- [ ]
- [ ]
- [ ]
- [ ]

Perch - External and internal anatomy

External anatomy

- ☐ Anal fin
- ☐ Anal opening
- ☐ Caudal fin
- ☐ Caudal peduncle
- ☐ Ctenoid scale
- ☐ Dermis
- ☐ Epidermis
- ☐ Excurrent naris (nostril)
- ☐ Eye
- ☐ First dorsal fin
- ☐ Head
- ☐ Incurrent naris (nostril)
- ☐ Lateral line
- ☐ Lips
- ☐ Lower mandible (dentary)
- ☐ Maxilla
- ☐ Mouth
- ☐ Mucus glands
- ☐ Operculum
- ☐ Pectoral fin
- ☐ Pelvic fin
- ☐ Premaxilla
- ☐ Second dorsal fin
- ☐ Tail
- ☐ Trunk
- ☐ Urogenital pore
- ☐
- ☐
- ☐
- ☐
- ☐

Muscular system

- ☐ Abductors of pectoral fin
- ☐ Abductors of pelvic fin
- ☐ Adductor caudalis
- ☐ Adductor mandibulae 1
- ☐ Adductor mandibulae 2
- ☐ Adductors of pelvic fin
- ☐ Dilator operculi
- ☐ Epaxial musculature
- ☐ Fin inclinator muscles
- ☐ Geniohyoideus anterior
- ☐ Geniohyoideus posterior
- ☐ Horizontal septum
- ☐ Hypaxial musculature
- ☐ Levator arcus palatini
- ☐ Levator hyoideus
- ☐ Levator operculi caudalis
- ☐ Levator operculi cranialis
- ☐ Myomeres
- ☐ Myosepta
- ☐ Rectus abdominis
- ☐ Sternohyoideus
- ☐ Trapezius
- ☐
- ☐
- ☐
- ☐
- ☐
- ☐
- ☐

Skeletal system

- ☐ Articula
- ☐ Basibranchials
- ☐ Basipterygium
- ☐ Branchiostegl rays
- ☐ Caudal vertebrae
- ☐ Centrum
- ☐ Ceratobranchial
- ☐ Ceratohyal
- ☐ Cleithrum
- ☐ Coracoid
- ☐ Dentary
- ☐ Dentigerous plate
- ☐ Epibranchial
- ☐ Epipleural rib
- ☐ Frontal
- ☐ Gill supports
- ☐ Hemal arch
- ☐ Hyobranchial
- ☐ Hyomandibular
- ☐ Hypohyal
- ☐ Interopercular
- ☐ Lacrimal
- ☐ Lepidotrichia
- ☐ Maxilla
- ☐ Metapterygoid
- ☐ Nasal
- ☐ Neural spine
- ☐ Opercular
- ☐ Pharyngobranchial
- ☐ Pleural rib
- ☐ Premaxilla
- ☐ Preopercular
- ☐ Prootic
- ☐ Pterygiophores
- ☐ Quadrate
- ☐ Radials
- ☐ Scapula
- ☐ Sphenotic
- ☐ Subopercular
- ☐ Suborbital bones
- ☐ Supracleithrum
- ☐ Symplectic
- ☐ Trunk vertebrae
- ☐ Urohyal
- ☐
- ☐
- ☐
- ☐

Internal Anatomy

- ☐ Afferent arteries
- ☐ Anterior cardinal vien
- ☐ Atrium
- ☐ Branchial arch
- ☐ Bulbus arteriosus
- ☐ Common cardinal vein
- ☐ Duodenum
- ☐ Esophagus
- ☐ Gall bladder
- ☐ Gill filaments
- ☐ Gill rakers
- ☐ Head kidney
- ☐ Heart
- ☐ Intestine
- ☐ Kidney
- ☐ Liver
- ☐ Ovary
- ☐ Pericardial sack
- ☐ Postcardinal vein
- ☐ Pyloric ceca
- ☐ Sinus venosus
- ☐ Spleen
- ☐ Stomach
- ☐ Swim bladder
- ☐ Testes
- ☐ Urinary bladder
- ☐ Ventral aorta
- ☐ Ventricle
- ☐
- ☐
- ☐

Crossword Puzzle - Jawed Fishes

Puzzle solution. An interactive web-based version of this puzzle and its solution are available on the *Digital Zoology* website www.digitalzoology.com/students. With the interactive puzzle you can check to see if individual words or the whole puzzle is correct and get hints for single letters.

Across

1   This moveable flap covering a fish's gills helps to pump water over the gills. (9)

3   Another name for a tail fin is this type of fin. (6)

5   Although you'll find these in invertebrate, these blind-ended extensions of a bony fish's stomach are specialized sites for digestion. (7,4)

12 In some fish a duct connects the swim bladder to this structure. (9)

14 The ventral unpaired fin in a shark. (4)

15 Number of gill filaments on each gill. (3)

16 This marvelous net is where gas enters and leaves the swim bladder. (4,7)

20 Although a shark doesn't have a large stomach it does have a special valve in the intestine that slows the passage of food and increases the surface area of the gut. (6)

21 These toothy scales are found in many bony fish. (7)

22 The posterior set of paired fins are these fins. (6)

23 The ampullae of Lorenzini in a cartilaginous fish are this type of receptor. (16)

25 These receptors are the sensory structure that detects vibrations that move through the water. (10)

27 The swim bladder ultimately becomes these in terrestrial vertebrates. (5)

29 There are no teeth on the edge of these almost circular scales in bony fish. (7)

30 A sharks teeth are highly modified versions of these surface structures. (6)

31 The fins on the left side of the body are connected to those on the right by these. (7)

32 You'll find these inside the dorsal lobe of a shark's tail. (9)

33 Sharks have four types of fins that help stabilize the animal as it swims. This is one of the unpaired fins. (6)

Down

2 A bony fish with neither pectoral nor pelvic fins. (3)

4 A shark's scales are formed from this layer of the integument. (6)

5 The most anterior pair of fins on the side of a shark are these fins. (8)

6 In a male shark the pectoral fin is modified with one of these that's used in mating. (7)

7 Bony fish are efficient at extracting oxygen from the water because of this; it describes how the two fluids flow past each other. (7,7)

8 Unlike the scales of a shark, those in bony fish never pierce this layer of the integument; instead the layer covers the scales. (9)

9 This oil helps keep a shark from sinking. (8)

10 The type of scales found only in cartilaginous fish. (7)

11 Bony fish maintain neutral density using this. (4,7)

13 The unique shape of a sharks tail, which has a dorsal lobe that is much larger than the ventral lobe. (12)

17 The platelike structures on the gill of a bony fish are where the gas exchange occurs. (8)

18 Another name for bony fish. (8)

19 The body shape of a fish; it's the best one for moving through a dense medium such as water. (8)

21 The skeleton of a shark is made of this material. (9)

24 True or false: Sharks lay eggs. (4)

25 Can a shark breath through its nostrils? (2)

26 These glands are found in the epithelial layer of a bony fish's integument. (5)

28 After leaving the heart of a fish, blood flows first to these structures. (5)

Self-Test - Jawed Fishes

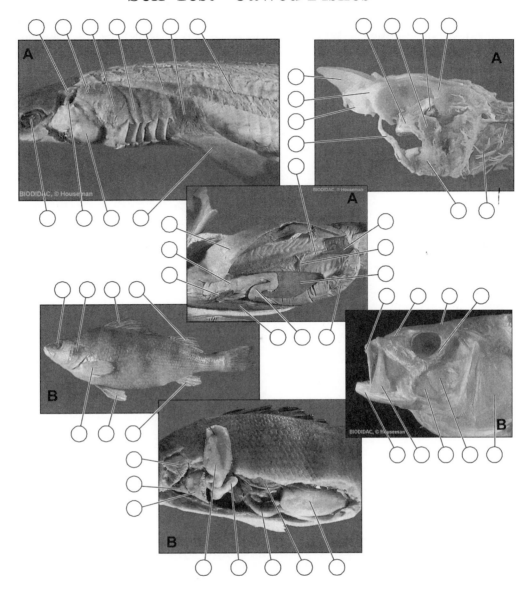

Use the following labels to identify the photographs. You may have to use a label more than once, and some labels may not be appropriate for the photographs. Answers are available on the *Digital Zoology* web site at www.digitalzoology.com/students. Be sure to try the interactive Drag-and-Drop quizzes that are available on the *Digital Zoology* CD-ROM.

Specimens

A- Dogfish shark
B- Perch

Labels

1) Anal fin
2) Antorbital process
3) Cardiac region of stomach
4) Cucullaris
5) Dentary
6) Eye
7) First dorsal fin
8) Frontal
9) Gall bladder
10) Gill filaments
11) Gill rays
12) Heart
13) Horizontal septum
14) Infraorbital bones
15) Intestine
16) Kidney
17) Labial cartilage
18) Levator hyomandibulae
19) Levator of pectoral fin
20) Liver

21) Mandibular adductor
22) Maxilla
23) Meckel's cartilage
24) Mesenteric artery
25) Metapterygoid
26) Nasal bone
27) Olfactory capsule
28) Operculum
29) Optic pedicle
30) Palatoquadrate cartilage
31) Pectoral fin
32) Pelvic fin
33) Peripharyngeal band (velum)
34) Pharynx
35) Postorbital process
36) Premaxilla
37) Proboscis coelom
38) Pyloric ceca
39) Pyloric region of stomach
40) Quadrate
41) Quadrate process
42) Rectal gland
43) Rostrum
44) Second dorsal constrictor
45) Second dorsal fin
46) Spiracle
47) Spiracular muscle
48) Spiral valve
49) Spleen
50) Stomach
51) Stomocord
52) Swim bladder
53) Testis
54) Third dorsal constrictor

AMPHIBIA

Inside *Digital Zoology*

As you explore the Amphibia on the *Digital Zoology* CD-ROM or on the web at www.digitalzoology.com, don't miss these learning tools:

- ○ Photos of dissected specimens of the mudpuppy and the frog including external anatomy and musculature, skeletal system, and internal anatomy.
- ○ Drag-and-drop quizzes on the external anatomy and musculature, skeletal system, and internal anatomy of the mudpuppy and the frog.
- ○ Interactive cladogram showing the major events that gave rise to vertebrate classes, and a summary of the key characteristics of each class are combined with an interactive glossary of terms.

Defining Differences

Some of the differences described in the following sections appear for the first time in the Amphibia, and they define the class. Others are important for understanding how these animals function. Whichever is the case, you'll want to watch for examples of these in the various amphibian specimens that you will be examining.

- ○ Transition to land;
- ○ tetrapod stance with a
- ○ strong, bony skeleton;
- ○ double circuit circulatory system; and a
- ○ moist, glandular skin.

Transition to land

Amphibians were the first vertebrates to move onto land. About 400 million years ago, during the Devonian Age, the earth was hot and dry. Aquatic environments often dried up or disappeared until flooding replenished them. As these bodies of water shrank, the density of fish and the other animals living in them increased. Crowding wasn't the only problem. The water became stagnant, depleted of oxygen. This was a problem for any animals that used **gills.** The solution was to breathe air and that required **lungs. Gas bladders** in fish form from outpockets of the pharynx or gut and have two functions in modern fish – respiratory gas exchange or a mechanism for achieving neutral buoyancy. It's not clear which came first in the ancient fish but a fish with a gas bladder capable of gas exchange connected to the alimentary tract and external nostrils could poke its head above water and breathe. There's fossil evidence to suggest that is what happened.

That solves the problem of poor oxygen quality. What about the increasing density of organisms living in the diminishing body of water? The solution was to get up and leave, and the strong fins of the lobe-finned fishes helped them do that. Their fin musculature and bones were a modification for crawling across the bottom while they fed, and they also worked for moving on land. The first fish that wiggled up on land wasn't out to conquer this new environment; it was trying to get to a better body of water. Ultimately this was the beginning of Amphibia, the first vertebrates to spend a significant amount of time on land.

Earth's environment was to undergo another radical change after the dry Devonian. The Carboniferous was a time when the earth was wet and warm; ideal for the new amphibians, and they diversified quickly. Coming on land was not only advantageous for escaping predators and a deteriorating environment but this was the time when insects started to conquer the terrestrial environment. Amphibians found a food source on which no one else was feeding.

There was one thing that amphibians couldn't do on land and that was reproduce. Their eggs still had to be laid in water and fertilized there. A swimming tadpole hatched from the egg and lived in the ancestral amphibian home until it was large enough to undergo **metamorphosis** and complete its life cycle on land.

Few species remain from the age of amphibians, and today's only representatives are salamanders, frogs, and a small group of caecilians. They represent the culmination of a long, diverse lineage of amphibians, and each is now specialized for its way of life. Neither of them looks like the supposed ancestral amphibian, and the same warning we made about cartilaginous and bony fish applies here. Don't think of salamanders as evolving into frogs. As we said with the two classes of fishes, it didn't happen that way.

Tetrapod stance

Amphibians were the first **tetrapods**. Their **appendicular skeleton** has undergone extensive modification, and now the **pectoral girdle** and **pelvic girdle** are firmly anchored to the **axial skeleton** using modified **vertebrae**. Attached to each of the girdles is a set of bones that form forelimbs and hindlimbs that hold the animal up on its four feet, the origins of the term tetrapod. This modification of the appendicular skeleton appears for the first time with the amphibians, and it's inherited by all the vertebrates that follow. The **taxon** is the Tetrapoda.

 How many different types of vertebrae are there in vertebrates?

Even though lung fish and lobe-finned fish may have had solutions to the problems of the Devonian, biologists still aren't certain which of the two was the one to come onto land. Whichever one it was, evolution of the tetrapod stance would follow the same sequence. The ancestral fish to the tetrapods strengthen its fins using **dermal bones,** called **fin rays**, anchored against a set of larger bones at their bases. The proposed changes required for the tetrapod stance saw a further strengthening of the bones that anchored the fin rays to the body and a decrease in the number of rays that ultimately became the **phalanges**. The first tetrapods probably had more than five phalanges on each limb but at some point the **pentatdactyl** plan appeared and all the vertebrates that followed inherited that plan.

In addition to the phalanges, what are the main bones of the tetrapod limbs?

Strong, bony skeleton

Movement on land required not only modifications of the pectoral and pelvic girdles, the **axial skeleton** had to change, too. As a skeletal element, the vertebral column of a fish had to resist compression along its length so that when the myomeres on each side contracted, the tail and trunk bent rather than shortened. Fish vertebrae are disc-shaped and embedded in the muscles of the tail and trunk. There was little chance of two vertebrae slipping by each other. That changed on land, and the vertebral column became a strengthening beam held up by the two sets of appendages. The appendages had to hook into the axial skeleton, which had to support the body. Vertebrae of tetrapods are considerably more complex than those of the fishes. Concave surfaces of the centrum fit into the convex surface of the adjacent vertebrae, and the neural arch is modified to form **zygapophyses**, which extend forward and backwards.

Double circuit circulatory system

The amphibian heart is more complex than that of the fishes. Instead of one **atrium,** there are now two – one that receives blood from the lungs and the other, blood that has circulated through the body. We now have two circulatory routes through the body, the **pulmonary circuit** and the **systemic circuit,** and a **dual circuit circulatory system**. With only one **ventricle** the separation of the two circuits isn't perfect, and there is some mixing of the oxygenated blood from the pulmonary path with the deoxygenated blood from the systemic system. This is kept to a minimum by slight differences in the timing of when blood from each of the atria enter the ventricle, the walls of the ventricle, and the **spiral valve** in the **conus arteriosus**. It's long been thought that the amphibian heart, with its single ventricle, was an intermediate step on the way to the four-chambered heart seen in some reptiles and all birds and mammals. This may not be the case. The amphibian ventricle is perfectly adapted to an animal that may only occasionally use its lungs. There's no point sending blood to the lungs if they aren't always being ventilated. If there was a permanent division of the amphibian ventricle, this is exactly what would happen each time the heart would beat.

✎   *Blood oxygenated in the skin passes into which part of the circulatory system?*

Moist, glandular skin

Although lungs were important in the transition from an aquatic to a terrestrial existence, **cutaneous respiration**, gas exchange across the skin, is still important for amphibians. It's the reason they have a moist, glandular skin not protected by scales that would interfere with gas exchange. The skin is covered by a **stratum corneum** formed from a few layers of dead epidermal cells filled with **keratin**. This new epidermal component is thick enough to protect against abrasion but it's not so thick that it would interfere with respiration. For any animal that breathes through the skin there needs to be a mechanism to keep the skin moist. In amphibians, large **mucus glands** do that. These aren't the only glands in an amphibian's skin. Granular **poison glands** secrete their contents and help protect against predators.

A Closer Look Inside *Digital Zoology* - Amphibia

Necturus: *The mudpuppy*

Of the three amphibian orders the most primitive are the salamanders, amphibians with tails. *Necturus* may not look like any salamander that you've seen before because it lives its whole life in water. It is fishlike in how it swims and breathes and terrestrial with its tetrapod stance and skeletal modifications associated with that. This makes the mudpuppy a good animal for looking at the transition from water to land. As you examine your specimen, or the one in *Digital Zoology*, you might notice another unusual characteristic about this animal. The adult breathes with larval gills.

The mudpuppy is an example of **neoteny**. It's a special case of **paedomorphosis**, which describes animals that retain their larval appearance but are sexually mature. A larval tunicate that underwent paedomorphosis is how Garstang proposed that the first chordates evolved. He didn't dream this up. There are biological precedents that he drew on, and the mudpuppy is one of them. In some ways the mudpuppy is an overgrown tadpole with legs and a reproductive system.

A mudpuppy's skeleton is composed of both **cartilaginous** and **bony** elements. The skull retains elements of the cartilaginous **chondrocranium,** which becomes covered by the dermal bones of the **dermatochranium** as the animal matures. A fish's head doesn't articulate with the vertebral column, so fish don't have necks. The mudpuppy doesn't either. Even though mudpuppies aren't terrestrial, their **vertebrae** have the overlapping processes, **prezagopophysis** and **postzygapophysis,** which strengthen the vertebral column so that it can be supported by the forelimb and hindlimb. The limbs aren't very big in a mudpuppy, and the pectoral girdle and pelvic girdle are a combination of cartilaginous and bony elements. Like cartilaginous fish, the pectoral girdle isn't attached to the vertebral column but is instead anchored in muscle. Only the pelvic girdle has **ilia** that connect it to the vertebral column as it does in terrestrial tetrapods.

The musculature is another good example of the transition to land with each girdle having its own set of muscles, which in salamanders are used to lift the body up and off the ground and to allow it to walk. But, like a fish, the mudpuppies axial musculature is still extensive with **epaxial** muscles positioned above a **horizontal septum** and **hypaxial** muscles below. The hypaxial musculature is the most extensive, and it's easy to see the three overlapping sets of **oblique muscles** below the horizontal septum.

Internally the urogenital system is fishlike and closely resembles that of the sharks. In the male, that anterior part of the **opisthonephric kidney** is involved in transporting sperm from the testes to the **archinephric duct** (mesonephric duct). The female urogenital system is also like a shark's, and eggs released from the **ovary** are swept into an **oviducal funnel** and the anterior of the **oviduct.**

Rana: *The grass frog*

The frog is a common dissection in zoology courses. It gives the opportunity to see some of the ancestral amphibian characters but also to see how the amphibian body has been modified for **saltatory locomotion**, jumping. The explosive muscular forces required for jumping have resulted in extensive modifications of both the axial and appendicular skeletons.

In jumping animals, the part of the leg that touches the ground is usually enlarged so that the forces generated from pushing up and off the ground can be spread over as much surface as possible. Frogs are no different, and their long toes help to increase the size of the hindfoot. Another modification that increases the size of the hindfoot are the two tarsal bones that have elongated to become the **astragulus (tibiale)** and **calcaneum (fubulare)**. A quick glance at the skeleton of a frog and you might mistake these two modified tarsal bones as the **tibia** and **fibula** because of their paired appearance. The real tibia and fibula have fused to form the **tibiofibula** bone, another modification that strengthens the hindlimb. In vertebrates, the **ilium** connects the pelvic girdle to the **sacral vertebrae,** and in frogs its length increases the size of the hindleg. There was an added advantage to this large foot that had nothing to do with jumping and that was for **natatorial locomotion**, swimming.

The axial skeleton has also been modified for jumping, and the **urostyle** is formed from the fusion of a number of postsacral vertebrae. The urostyle is positioned between the two large ilial bones. Muscles between the two help to strengthen the connection between the appendicular and axial skeletons. Even the forelimbs are modified for jumping, or more precisely, landing. When a frog jumps, it lands with its front legs first. The radius and ulna have fused into a stronger **radioulna**.

Cross-References to Other McGraw-Hill Zoology Titles

Integrated Principles of Zoology, 11th edition. C. P. Hickman, L. S. Roberts, and A. Larson. Chapter 27.

Animal Diversity, 3rd edition. C. P. Hickman, L. S. Roberts, and A. Larson. Chapter 17.

Zoology, 5th edition. S. A. Miller and J. P. Harley. Chapter 19.

Vertebrate Biology, 1st edition. D. W. Linzey. Chapter 6.

Laboratory Studies in Integrated Principles of Zoology, 10th edition. C. P. Hickman, F. Hickman, and L. Kats. Chapter 19.

General Zoology Laboratory, 13th edition. C. Lytle. Chapter 18.

Structures Checklist

Here are some of the structures that you should be able to easily find in *Digital Zoology* and the specimens that you will be looking at in your lab. After reading your lab handout, you might want to add more and, depending on the equipment available in your lab, you might see more. As you study the material, you might also want to make some notes on how some of these structures looked or include a drawing in your lab notes. (Structures indicated by * may be hard to see.)

Mudpuppy - External anatomy and skeletomuscular systems

External anatomy

- [ ] Chromatophores
- [ ] Cloacal opening
- [ ] Dermis
- [ ] Epidermis
- [ ] External gills
- [ ] External naris (nostrils)
- [ ] Eyes
- [ ] Gular fold
- [ ] Head
- [ ] Lips
- [ ] Mouth
- [ ] Mucus glands
- [ ] Pectoral limb
- [ ] Pelvic limb
- [ ] Poison glands
- [ ] Tail
- [ ] Trunk
- [ ]
- [ ]
- [ ]
- [ ]
- [ ]
- [ ]

Musculature - Head and trunk

- [ ] Anterior mandibular levator
- [ ] Branchial levators
- [ ] Branchiohyoid
- [ ] Dorsalis trunci
- [ ] Epaxial musculature
- [ ] Eternal oblique

- [ ] External mandibular levator
- [ ] Geniohyoid
- [ ] Horizontal septum
- [ ] Hypaxial musculature
- [ ] Interhyoid
- [ ] Intermandibular
- [ ] Internal oblique
- [ ] Linea alba
- [ ] Mandibular depressor
- [ ] Myomeres
- [ ] Myosepta
- [ ] Rectus abdominus
- [ ] Rectus cervicis
- [ ] Transversus abdominus
- [ ]
- [ ]
- [ ]

Musculature - Appendages

- [ ] Caudofemoralis
- [ ] Caudopuboischio-tibilias (caudocruralis)
- [ ] Coracobrachialis
- [ ] Cucullaris
- [ ] Dorsal scapulae
- [ ] Humeroantebrach-ialis
- [ ] Ilioextensprius
- [ ] Iliofubularis

- [ ] Iliotibialis
- [ ] Ishioflexorius
- [ ] Latissimus dorsi
- [ ] Pectoralis
- [ ] Pectoriscapularis
- [ ] Procoracohumeralis
- [ ] Pubioischiofemoralis internus
- [ ] Puboischiofemoralis externus
- [ ] Puboischiotibialis
- [ ] Pubotibialis
- [ ] Supracotacoideus
- [ ] Triceps
- [ ]
- [ ]
- [ ]
- [ ]

Skeleton - Skull and trunk

- [ ] Angula
- [ ] Basibranchials
- [ ] Capitulum
- [ ] Caudal vertebrae
- [ ] Centrum
- [ ] Ceratobranchials
- [ ] Ceratohyal
- [ ] Cervical vertebra (atlas)
- [ ] Columella
- [ ] Dentray
- [ ] Diapophysis
- [ ] Epibranchials
- [ ] Ethmoid

- [ ] Foramen magnum
- [ ] Frontals
- [ ] Hyoid arch
- [ ] Hypohyal
- [ ] Meckel's cartilage
- [ ] Neural arch
- [ ] Neural canal
- [ ] Neural spine
- [ ] Occiptal condyle
- [ ] Opisthotic
- [ ] Palatopterygoids
- [ ] Parapophysis
- [ ] Parietals
- [ ] Postzygapophysis
- [ ] Premaxilla
- [ ] Prezygapophysis
- [ ] Prootic
- [ ] Quadrates
- [ ] Rib
- [ ] Sacral vertebra
- [ ] Splenial
- [ ] Squamosals
- [ ] Transverse process
- [ ] Trunk vertebrae
- [ ] Tuberculum
- [ ] Visceral arches
- [ ] Vomers
- [ ]
- [ ]
- [ ]
- [ ]

Mudpuppy - Skeletal (cont.), and internal systems

Skeleton - Appendages

- ☐ Acetabulum
- ☐ Carpals
- ☐ Coracoid cartilage
- ☐ Femur
- ☐ Fibula
- ☐ Glenoid fossa
- ☐ Humerus
- ☐ Ilium
- ☐ Ischium
- ☐ Metacarpals
- ☐ Phalanges
- ☐ Procoracoid cartilage
- ☐ Pubis
- ☐ Puboishiadic plate
- ☐ Radius
- ☐ Scapula
- ☐ Suprascapular cartilage
- ☐ Tibia
- ☐ Ulna
- ☐
- ☐
- ☐
- ☐
- ☐
- ☐

Digestive and related systems

- ☐ Dentary teeth
- ☐ Duodenum
- ☐ Esophagus
- ☐ Glottis
- ☐ Ileum
- ☐ Liver
- ☐ Mouth
- ☐ Pancreas
- ☐ Pharynx
- ☐ Premazillary teeth
- ☐ Pterygoid teeth
- ☐ Rectum
- ☐ Spleen
- ☐ Splenial teeth
- ☐ Stomach
- ☐ Vomerine teeth
- ☐
- ☐
- ☐
- ☐

Circulatory and respiratory systems

- ☐ Afferent branchial arteries
- ☐ Anterior mesenteric artery
- ☐ Atria
- ☐ Axillary vein
- ☐ Brachial vein
- ☐ Bulbus arteriosus
- ☐ Celiac artery
- ☐ Celiacomesenteric artery
- ☐ Cloacal arteries
- ☐ Common cardinal veins
- ☐ Conus arteriosus
- ☐ Cutaneous vein
- ☐ Dorsal aorta
- ☐ Efferent branchial arteries
- ☐ External carotid artery
- ☐ External jugular vein
- ☐ Gastric artery
- ☐ Gastrosplenic vein
- ☐ Genital arteries
- ☐ Hepatic artery
- ☐ Hepatic portal vein
- ☐ Iliac arteries
- ☐ Internal carotid artery
- ☐ Internal jugular vein
- ☐ Intestinal arteries
- ☐ Lung
- ☐ Mesenteric vein
- ☐ Pancreaticoduodenal artery
- ☐ Posterior cardinal vein
- ☐ Posterior mesenteric artery
- ☐ Posterior vena cava
- ☐ Pulmonary artery
- ☐ Pulmonary vein
- ☐ Radix aorta
- ☐ Renal arteries
- ☐ Sinus venosus
- ☐ Splanic artery
- ☐ Subclavian artery
- ☐ Subclavian vein
- ☐ Ventral abdominal vein
- ☐ Ventricle
- ☐ Vertebral artery
- ☐
- ☐
- ☐
- ☐
- ☐
- ☐
- ☐

Urogenital systems

- ☐ Cloaca
- ☐ Efferent ductules
- ☐ Kidneys
- ☐ Mesonephric ducts (Archinephric duct)
- ☐ Mesorchium
- ☐ Mesotubarium
- ☐ Mesovarium
- ☐ Ostium (oviduct)
- ☐ Ovaries
- ☐ Testes
- ☐ Urinary bladder
- ☐
- ☐
- ☐
- ☐
- ☐
- ☐

Additional structures

- ☐
- ☐
- ☐
- ☐
- ☐
- ☐
- ☐
- ☐
- ☐
- ☐

Frog - Extrenal anatomy and skeletomuscular systems

External anatomy

- [ ] Ankle
- [ ] Antebrachium
- [ ] Brachium
- [ ] Chromophores
- [ ] Cloacal opening
- [ ] Crus (leg)
- [ ] Dermis
- [ ] Digits (fingers)
- [ ] Elbow
- [ ] Epidermis
- [ ] External nares (nostrils)
- [ ] Eye
- [ ] Eye lids
- [ ] Femur (thigh)
- [ ] Head
- [ ] Knee
- [ ] Manus (hand)
- [ ] Mouth
- [ ] mucus glands
- [ ] Nictitating membrane
- [ ] Pes (foot)
- [ ] Poison glands
- [ ] Pollex (thumb)
- [ ] Tibiotarsalus
- [ ] Trunk
- [ ] Tympanic membrane
- [ ]
- [ ]
- [ ]
- [ ]

Musculature - Head and trunk

- [ ] Coccygeoiliacus
- [ ] Cutaneous abdominis
- [ ] Cutaneous pectoris
- [ ] Depressor mandibulae
- [ ] Dorsalis scapulae
- [ ] External oblique
- [ ] Geniohyoideus
- [ ] Iliolumbaris
- [ ] Internal oblique
- [ ] Latissimus dorsi
- [ ] Longissimus dorsi
- [ ] Masseter
- [ ] Mylohyoid (submaxillary)
- [ ] Omohyoideus
- [ ] Petrohyoideus
- [ ] Pterygoideus
- [ ] Rectus abdominis
- [ ] Rhomboideus
- [ ] Sternohyoideus
- [ ] Subhyoid
- [ ] Temporalis
- [ ] Transverse oblique
- [ ]
- [ ]
- [ ]
- [ ]

Musculature - Appendages

- [ ] Abductor indicis longus
- [ ] Adductor longus
- [ ] Adductor magnas
- [ ] Coracobranchialis
- [ ] Crualis (vastus internus)
- [ ] Cucullaris
- [ ] Deltoideus
- [ ] Dorsalis scapulae
- [ ] Extensor carpi radialis
- [ ] Extensor carpi ulnaris
- [ ] Extensor cruris brevis
- [ ] Extensor digitorum communis longus
- [ ] Flexor carpi radialis
- [ ] Flexor carpi ulnaris
- [ ] Flexor digitorum
- [ ] Gastrocnemius
- [ ] Gluteus (iliacus externus)
- [ ] Gluteus magnus (vastus externus)
- [ ] Gracilis major (rectus internus major)
- [ ] Gracilis minor (rectus internus minor)
- [ ] Iliacus internus
- [ ] Iliofemorallis
- [ ] Iliofubularis (biceps femoris)
- [ ] Latissimus dorsi
- [ ] Palmaris longus
- [ ] Pectineus
- [ ] Pectoralis major
- [ ] Pectoralis minor
- [ ] Peroneus
- [ ] Plantaris longus
- [ ] Quadratus femoris
- [ ] Sartorius
- [ ] Semimembranosus
- [ ] Semitendinosus
- [ ] Tarsalis anterior
- [ ] Tarsalis posterior
- [ ] Tendon of Achilles
- [ ] Tensor fasciae latae
- [ ] Tibialis anticus longus
- [ ] Tibialis posticus longus
- [ ] Triceps brachii (Anconeus)
- [ ] Triceps femoris
- [ ]
- [ ]
- [ ]
- [ ]

Skeleton - Skull and trunk

- [ ] Articular process (zygopophyses)
- [ ] Atlas
- [ ] Centrum
- [ ] Columella
- [ ] Dentary
- [ ] Exoccipital
- [ ] Foramen magnum
- [ ] Frontoparietal
- [ ] Maxilla
- [ ] Nasal
- [ ] Neural arch
- [ ] Neural spine
- [ ] Occiptal condyle
- [ ] Orbit
- [ ] Otic capsule
- [ ] Palatine
- [ ] Parasphenoid
- [ ] Premaxilla
- [ ] Prootic
- [ ] Pterygoid
- [ ] Quadrate
- [ ] Quadratojugal
- [ ] Sacral vertebrae
- [ ] Sphenethmoid
- [ ] Squamosal
- [ ] Sternum
- [ ] Supraoccipital
- [ ] Transverse process
- [ ] Trunk vertebrae
- [ ] Urostyle
- [ ] Vomer
- [ ]
- [ ]

Frog - Skeletal (cont.), and internal systems

Skeleton - appendages

- ☐ Acetabulum
- ☐ Astragalus (tarsus)
- ☐ Calcaneus (tarsus)
- ☐ Carpals
- ☐ Clavicle
- ☐ Coracoid
- ☐ Femur
- ☐ Glenoid fossa
- ☐ Humerus
- ☐ Ilium
- ☐ Ischium
- ☐ Metacarpals
- ☐ Metatarsals
- ☐ Phalanges
- ☐ Pubis
- ☐ Radioulna
- ☐ Scapula
- ☐ Sternum
- ☐ Suprascapula
- ☐ Tarsals
- ☐ Tubiofibula
- ☐
- ☐
- ☐
- ☐
- ☐

Digestive and related systems

- ☐ Cloaca
- ☐ Colon
- ☐ Common bile duct
- ☐ Cystic duct
- ☐ Duodenum
- ☐ Esophagus
- ☐ Eustachian tube
- ☐ Gall bladder
- ☐ Glottis
- ☐ Internal nares
- ☐ Jejunoileum
- ☐ Liver
- ☐ Mesenteries
- ☐ Pancreas
- ☐ Pharynx
- ☐ Pyloric sphincter
- ☐ Spleen
- ☐ Stomach
- ☐ Tongue
- ☐ Vocal sacs
- ☐ Vomerine teeth
- ☐
- ☐
- ☐
- ☐

Circulatory and respiratory systems

- ☐ Anterior mesenteric artery
- ☐ Anterior vena cava
- ☐ Brachial artery
- ☐ Brachial vein
- ☐ Celiacomesenteric artery
- ☐ Celic artery
- ☐ Common carotid artery
- ☐ Conus arteriosus
- ☐ Cutaneous artery
- ☐ Dorsal aorta
- ☐ Dorsolumbar vein
- ☐ External carotid artery
- ☐ External jugular
- ☐ Femoral artery
- ☐ Femoral vein
- ☐ Gastric artery
- ☐ Gastric vein
- ☐ Gonadial vein
- ☐ Heart
- ☐ Hepatic artery
- ☐ Hepatic portal vein
- ☐ Hepatic vein
- ☐ Iliac artery
- ☐ Iliac vein
- ☐ Innominate vein
- ☐ Internal carotid artery
- ☐ Internal jugular vein
- ☐ Left atrium
- ☐ Lingual vein
- ☐ Lumbar arteries
- ☐ Lung
- ☐ Mandibular vein
- ☐ Mesenteric veins
- ☐ Musculocutaneous vein
- ☐ Pelvic vein
- ☐ Posterior mesenteric artery
- ☐ Posterior vena cava
- ☐ Pulmocutaneous arch
- ☐ Pulmonary artery
- ☐ Pulmonary vein
- ☐ Renal artery
- ☐ Renal portal vein
- ☐ Renal vein
- ☐ Right atrium
- ☐ Sciatic artery
- ☐ Sciatic vein
- ☐ Sinus venosus
- ☐ Splenic artery
- ☐ Splenic vein
- ☐ Subclavian artery
- ☐ Subclavian vein
- ☐ Subscapular vein
- ☐ Systemic arch
- ☐ Truncus arteriosus
- ☐ Urogenital arteries
- ☐ Ventral abdominal vein
- ☐ Ventricle
- ☐
- ☐
- ☐
- ☐
- ☐

Urogenital systems

- ☐ Adrenal glands
- ☐ Archinephric duct (Wolffian duct)
- ☐ Efferent ductules
- ☐ Fat body
- ☐ Kidney
- ☐ Mesorchium
- ☐ Mesovarium
- ☐ Ostium
- ☐ Oviduct
- ☐ Seminal vesicle
- ☐ Testes
- ☐ Ureters
- ☐ Urinary bladder
- ☐ Uterus
- ☐
- ☐
- ☐
- ☐
- ☐

Additional structures

- ☐ Subcutaneous lymph sacs
- ☐
- ☐
- ☐
- ☐

Crossword Puzzle - Amphibia

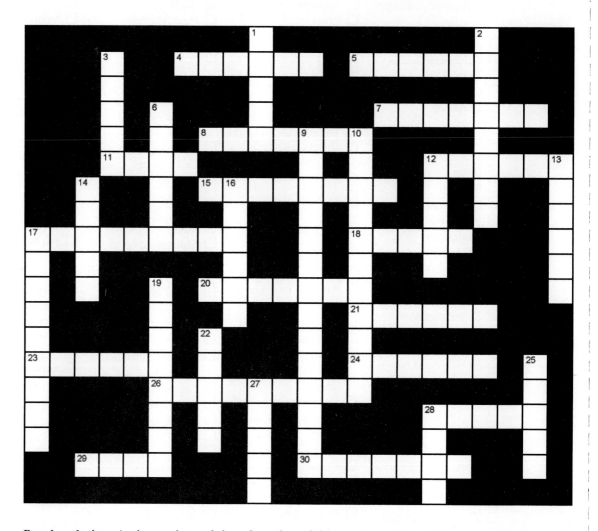

Puzzle solution. An interactive web-based version of this puzzle and its solution are available on the *Digital Zoology* website www.digitalzoology.com/students. With the interactive puzzle you can check to see if individual words or the whole puzzle is correct and get hints for single letters.

Across

4 This and the tibia form the lower bones of the tetrapod hindlimb. (6)

5 A ligular flap refers to this structure in a frog. (6)

7 An immature larval amphibian. (7)

8 Probable food for the first amphibians on land. (7)

11 The main respiratory surface in most amphibians. (4)

12 This valve in the ventricle of an amphibian heart separates blood flow from the parts of the circulatory system. (6)

15 An amphibian can breathe through these when its mouth is shut. (8)

17 The extra loop in the amphibian circulatory system missing in fish. (9)

18 Lungs, skin, and the inner surface of this can be used to varying degrees in amphibians for gas exchange. (5)

20 Outer cells of the amphibian epidermis have this to strengthen them, but not so much that gas exchange is inhibited. (7)

21 Describes a circulatory system with two routes that blood can take as it passes through the heart. (6)

23 To be able to keep amphibian skin moist you'll find many of these in the skin. (6)

24 This and the ulna form the lower bones of the tetrapod forearm. (6)

26 The presence of fore- and hindlimbs identifies amphibians and all the vertebrates to follow as this group of animals. (9)

28 The single upper bone in the tetrapod hindlimb. (5)

29 The tetrapod limb is pentadactyl and has this many fingers or toes. (4)

30 This bone, an addition to the axial skeleton of a frog, helps support and contain the internal organs. (7)

Down

1 Limbs were one of the things advantageous for moving on land; this was one of the others. (5)

2 During this geological time the earth was warm and dry and there was an advantage to be able to move between bodies of water. It's one of the reasons that amphibians appeared. (8)

3 These glands in an amphibian help to keep its skin moist. (5)

6 These glands in an amphibian are defensive. (6)

9 These gives an amphibian's skin its colors. (14)

10 Frogs, caecillans, and these are the only living amphibians. (11)

12 In amphibians blood flow to the lungs comes from this aortic arch. (5)

13 The amphibian voice box. (6)

14 Many amphibians demonstrate paedomorphosis and these appear in the adult. (5)

16 Compared to water, there is more of this in air. (6)

17 The transition from water to air appears in the amphibian's life cycle, ontogeny, as well as this history of the group. (9)

19 In a frog these are formed from fused vertebrae and are an adaptation related to jumping. (8)

22 To survive, all amphibians have to lay their eggs in this. (5)

25 Number of chambers in an amphibian heart. (5)

27 With the movement on land, this type of musculature isn't as important in amphibians as it was in fishes. (5)

28 The number of limbs that tetrapod refers to. (4)

Self-Test - Amphibia

Use the following labels to identify the photographs. You may have to use a label more than once, and some labels may not be appropriate for the photographs. Answers are available on the *Digital Zoology* website at www.digitalzoology.com/students. Be sure to try the interactive drag-and-drop quizzes on the *Digital Zoology* CD-ROM.

Specimens

A- *Rana* - Frog
B- *Necturus* -
   Mudpuppy

Labels

1) Conus arteriosus
2) Cucullaris
3) Depressor mandibulae
4) Dilator laryngis
5) Dorsalis scapulae

6) Duodenum
7) Eye
8) Femur
9) Fibula
10) Gall bladder
11) Gill
12) Hepatic vein
13) Ileum

14) Ilium
15) Interhyoideus
16) Ischium
17) Latissimus dorsi
18) Levatores arcuum
19) Levator mandibulae anterior

20) Levator mandibulae externus
21) Liver
22) Lung
23) Mesentery
24) Metatarsals
25) Ovary
26) Oviduct
27) Phalange
28) Pterygoideus
29) Pubis
30) Radioulna
31) Rhomboideus
32) Right atrium of heart
33) Small intestine
34) Spleen
35) Stomach
36) Suprascapula
37) Tarsus calcaneum
38) Temporalis
39) Tibia
40) Tibiofibula
41) Truncus arteriosus
42) Tympanum
43) Urostyle
44) Ventral abdominal vein
45) Ventricle of heart
46) Vertebrae

MAMMALIA

Inside *Digital Zoology*

As you explore the Mammalia on the *Digital Zoology* CD-ROM or on the web at www.digitalzoology.com, don't miss these learning tools:

○ Photos of dissected specimens of the fetal pig including external anatomy and musculature, skeletal system, and internal anatomy.

○ Drag-and-drop quizzes on the external anatomy and musculature, skeleton, and internal anatomy of the fetal pig.

○ Interactive cladogram showing the major events that gave rise to vertebrate classes, and a summary of the key characteristics of each class are combined with an interactive glossary of terms.

Defining Differences

Some of the differences described in the following sections appear for the first time in the Mammalia, and they define the class. Others are important for understanding how these animals function. Whichever is the case, you'll want to watch for examples of these in the specimens that you will be examining.

○ Homeothermic endotherms with
○ hair,
○ glandular skin,
○ heterodent diphyodont dentition, a
○ muscular diaphragm, and
○ extended care of young.

Homeothermic endotherms

Mammals and birds are the only two taxa of animals that maintain their bodies at a precise and constant temperature, **homeothermy,** and they generate heat to maintain that temperature, **endothermy**. All of life's processes are based on biochemical reactions, and reaction rates are temperature dependent. The lower the temperature, the slower the reaction. Raise the temperature and the reactions speed up only to a point that the enzymes start to breakdown from the heat. There is an advantage to guaranteeing that life's biochemical reactions happen at the same temperature, and homeothermic endotherms have this advantage.

For an animal to maintain a constant body temperature there needs to be a way to generate heat when it's necessary and a way to get rid of it if the animal gets too hot. Heat is a normal by-product of the biochemical reactions of metabolism and is used by mammals and birds as a source of heat. **Shivering** is muscle contractions producing metabolic heat as a by-product of the contractions. Oxidative metabolism of brown fats is another source of metabolic heat. A homeotherm can't spend its time generating heat, and there's an advantage to retaining heat by insulating the body. The fur coat is a mammal's insulative covering.

How does countercurrent blood flow in the extremities minimize cooling of the body core of a mammal?

That takes care of generating heat; what about cooling? A mammal's temperature is lowered by evaporation of watery sweat poured onto the surface of the skin by sweat glands. Sweat glands aren't the only way that mammals use evaporative cooling. The nasal passages or tongue are other important surfaces that mammals may use to lower their body temperature.

Hair

Just as feathers are unique to birds, **hair** is only found on mammals. Mammalian skin has the same two layers found in all vertebrates, an outer **epidermis** and an underlying **dermis**. The epidermis is much thicker than the dermis and is differentiated into a number of layers. New epidermal cells form in the bottom **stratum germinativum,** and as newer cells form, they push the older cells above them to the surface. As this happens, cells become **keratinized,** and the outermost layer of the skin, **stratum corneum,** is formed from dead keratinized cells, which waterproof mammalian skin and protect it against abrasion.

The **hair follicle** is a fold in the lower layers of the epidermis that penetrates deep into the dermal part of the skin. The dermis forms the **dermal papillae** at the base of the hair, and it's surrounded by the **epidermal hair matrix** formed from the stratum germinativum. It's the matrix that produces the hair. Mitotic cell divisions add new living cells at the hairs' base. As the new cells are added, the old keratinized cells of the hair die and are pushed up and out of the follicle. The result is a hair shaft composed of a **central cortex** surrounded by a **medula** completely wrapped inside overlapping **scale cells**.

Most mammals are covered with **fur**, or **pelage**, formed from two types of hairs. Longer, tough **guard hairs** protect the softer insulating **underfur** found next to the body. Guard hairs can also be modified into **quills** in animals such as the porcupine, or tactile **vibrissae**, whiskers found around the snouts of many mammals. Hairs usually lie at an angle to the skin and point backward. Contraction of the **arrector pilli muscle** causes the hair to rise to a more vertical position and increases the layer of insulating air against the skin. When you get goosebumps, it's the arrector pilli at work.

Glandular skin

The various glands in mammalian skin are one of two types: **sebaceous glands** or **sweat glands**. Hair is composed of dead cells, and if it were not "conditioned" in some way it would become brittle, break, or loose its waterproofing properties. To prevent this, sebaceous glands secrete oils and waxes, **sebum,** into the follicle, and the hair acts like a wick drawing up the secreted sebum, hairs' "natural" conditioner.

As their name implies, sweat glands produce watery sweat squeezed to the surface by the myoepithelium, which lines the ducts of the gland. Sweat glands are usually restricted to specific locations on a mammal's body and that location depends on the species. **Eccrine sweat glands** produce a watery fluid secreted onto the surface of the skin, and as it evaporates, it cools the body. **Apocrine sweat glands** look the same as eccrine glands but they release a thicker and often scented fluid into the hair follicle instead of the skin surface. The scents and smells produced by apocrine glands are important for communications in mammals.

The last type of gland gives the class Mammalia their name, **mammary glands**. The structure of the mammary gland resembles sebaceous glands but they have a myoepethelium like sweat glands. Whatever their origin, the glands secrete a nutritive fluid of fats, carbohydrates, and proteins used to feed the young. In the most primitive mammals, the monotremes, the mammary glands are grouped in a milk patch on the surface of the skin. To feed, a young monotreme licks the skin surface of the parent. Both male and female monotremes have functional mammary glands! Teats or nipples can be found in all other mammals and they form a long milk ridge that runs down the ventral side of the developing fetus. How many depends on the animal and the number of young that it produces.

Heterodent diphyodont dentition

In early vertebrates teeth were uniform, cone-shaped structures embedded in the jaw. Mammals have a different type of dentition and specialized teeth with different functions. The presence of **incisors**, **canines**, **premolars,** and **molars** usually matches the food and feeding strategy of the mammal.

✍ *For what is each type of tooth modified?*

Mammals also have not one, but two sets of teeth during their lives. This is the **diphyodont** condition. The first are the milk teeth, and they will be replaced by the permanent teeth. As the animal grows, the jaw enlarges. When the second set of teeth are added, the jaw has finished growing. Molars only appear with the permanent teeth.

Muscular diaphragm

In mammals the floor of the thoracic cavity is lined with a muscular **diaphragm,** which, at rest, is dome-shaped. When the muscles in the diaphragm contract, it flattens, increasing the size of the thoracic cavity. At the same time, the **intercostal muscles** raise the rib cage, increasing the size of the thoracic cavity. The lungs are suspended in the cavity and the negative pressure created by the expanding thoracic cavity pulls air into them. Once inside the lungs, gas exchange occurs in the **alveoli,** which increase the available surface area for gas exchange. The **dual circuit circulatory system** is complete in mammals, and the pulmonary and systemic circuits are completely separated in the heart.

Extended care of young

The reproductive strategies of animals can be represented by two extremes. One is to produce huge numbers of eggs and hope that a few of them will survive to be reproductive adults. The opposite strategy is to produce only a few young and help them on their way to adult maturity. The first is the **r-selection** strategy and the second is the **K-selection** strategy. The thousands of eggs that a female insect lays is a good example of the r-selection strategy. Mammals use the K-selection strategy. Mammalian females produce only a few young during each reproductive cycle. Their young are protected internally as they grow and after their birth stay with the parent and are fed using the mammary glands. During **suckling**, and **weaning** that follows, young mammals are protected and learn a variety of behaviors and habits important for their survival. The lengths of time that the young are cared for, and how mature they are when born, varies in different mammals.

What may surprise you is that not all mammals are **viviparous**. The most primitive are the **oviparous** monotremes that lay eggs. Marsupials go a step farther, and the young embryo develops in the uterus for a short time before crawling out and into a protective pouch where the mammary glands are found. The last group of mammals are the placental mammals, and the developing embryo uses the **placenta** to exchange nutrients, oxygen, and metabolic wastes with the mother.

A Closer Look Inside *Digital Zoology* - Mammalia

The fetal pig

Mammals, like birds, have completely separated the pulmonary and systemic circuits of the circulatory system. Oxygen-poor blood that has circulated through the body enters the right atrium of the heart through the **caudal vena cava** and **cranial vena cava** from the lower and upper parts of the body, respectively. From there it goes to the right ventricle, which pumps it into the **pulmonary artery** and on to the lungs for oxygenation. The oxygen-rich blood returns to the heart's left atrium through the **pulmonary vein**, passes to the left ventricle, and is then pumped out to the body through the **aortic arch**. The first major branch to come off the aortic arch is the **brachiocephalic artery,** which forms the **right subclavian artery** and the **common carotid artery,** which will branch to form the left and right common carotids that branch even farther as they supply the head. The second branch to come off the aortic arch is the **left subclavian artery** completing the pair of arteries that supply the anterior part of the body and the forelimbs. The aortic arch forms the **dorsal aorta** that runs the length of the animal, and its main branches supply different organs and parts of the body: the stomach, pancreas, and spleen, with the **celiac artery**; the liver, the **hepatic artery**; the mesenteries, the cranial and caudal **mesenteric arteries**; the adrenal gland, the **adrenal arteries**; the kidneys, the **renal arteries**. **Genital arteries** branch off the dorsal vessel to supply gonads, and the posterior end of the dorsal aorta branches to form the **common iliac artery,** which supplies blood to the hindlimbs.

In most cases, blood from the lower part of the body returns to the heart through a set of veins corresponding to the arteries that supplied the region. It's a little more complicated in the anterior region where external and internal **jugular veins** from the head join with the **cepahlic veins** and **subclavian veins** on each side to form the left and right **branchiocephalic veins**. These then fuse to form the cranial vena cava, which joins the caudal vena cava at the right atrium of the heart. The digestive system doesn't return its blood directly to the caudal vena cava like other organs do. Instead, blood from the alimentary tract passes to the liver through the **hepatic portal system**. Only after it has passed through the capillary beds of the liver does it return to the caudal vena cava on its way back to the heart.

 What is the role of the hepatic portal system?

At some point during embryonic development each of the original six aortic arches makes their appearance, three will disappear during development, and only three will remain when the mammal is born: the **third aortic arches** forming the internal carotid arteries. The large aortic arch is formed from the left side of the **fourth aortic arch** and the right hand side forms the subclavian artery. Ancestrally both of these aortic arches would have connected with the dorsal aorta but in mammals only the left side makes the connection. The **sixth aortic arches** supply the lungs.

A fetus' lungs aren't functional until after it's born, and there's no point sending large volumes of blood through the pulmonary circuit prior to that. The **ductus ateriosus,** which connects the pulmonary artery to the dorsal aorta, short circuits the system. Blood that would otherwise have gone to the lungs goes to the dorsal aorta and into the systemic circuit. This is the first of two short circuits in the fetal circulatory system. The second is inside the heart. The **foramen ovale**, an opening between the right and left atriums, allows blood returning from the body that enters the right atrium, to flow into the left atrium and from there to the left ventricle and back out to the body.

A fetus depends entirely on the placental blood supply for oxygen. An umbilical vein from the placenta connects to the fetal circulatory system through the **ductus venosus,** which passes blood to the liver and the caudal vena cava. From there it goes directly to the fetal heart. Blood returns to the placenta through the umbilical arteries connected to the posterior end of the dorsal aorta. When the young mammal is born, placental circulation stops, and with its first breath, a muscle inside the ductus arteriosis closes that vessel and fibrous tissue seals the tube forming a ligament that connects the dorsal aorta and pulmonary artery. Muscles also contract and close the umbilical arteries, which form ligaments or the common and internal iliac arteries.

Cross-References to Other McGraw-Hill Zoology Titles

Integrated Principles of Zoology, 11th edition. C. P. Hickman, L. S. Roberts, and A. Larson. Chapter 30.

Animal Diversity, 3rd edition. C. P. Hickman, L. S. Roberts, and A. Larson. Chapter 20.

Zoology, 5th edition. S. A. Miller and J. P. Harley. Chapter 22.

Vertebrate Biology, 1st edition. D. W. Linzey. Chapter 9.

Laboratory Studies in Integrated Principles of Zoology, 10th edition. C. P. Hickman, F. Hickman, and L. Kats. Chapter 22.

General Zoology Laboratory Guide, 13th edition. C. Lytle. Chapters 19 & 20.

Structures Checklist

Here are some of the structures that you should be able to easily find in *Digital Zoology* and the specimens that you will be looking at in your lab. After reading your lab handout, you might want to add more and, depending on the equipment available in your lab, you might see more. As you study the material, you might also want to make some notes on how some of these structures looked or include a drawing in your lab notes. (Structures indicated by * may be hard to see.)

Fetal pig - External anatomy and muscular systems

External anatomy

- [ ] Ankle
- [ ] Anus
- [ ] Auricle (ear)
- [ ] Digits
- [ ] Elbow
- [ ] External nares (nostrils)
- [ ] Eyelids
- [ ] Eyes
- [ ] Genital papilla
- [ ] Head
- [ ] Hooves
- [ ] Knee
- [ ] Mammary papilla
- [ ] Neck
- [ ] Nictitating membrane
- [ ] Scrotal sacs
- [ ] Shoulder
- [ ] Tongue
- [ ] Trunk
- [ ] Umbilical cord
- [ ] Urogenital opening
- [ ] Vibrissae
- [ ] Wrist
- [ ]
- [ ]
- [ ]
- [ ]
- [ ]

Musculature - Head and trunk

- [ ] Brachiocephalicus
- [ ] Digastric
- [ ] External oblique
- [ ] Internal oblique
- [ ] Linea alba
- [ ] Masseter
- [ ] Mylohyoid
- [ ] Omohyoid
- [ ] Rectus abdominis
- [ ] Rectus thoracis
- [ ] Sternohyoid
- [ ] Sternomastoid
- [ ] Sternothyroid
- [ ] Temporalis
- [ ] Thyrohyoid
- [ ]
- [ ]
- [ ]
- [ ]

Musculature - Pectoral and forelimb

- [ ] Biceps
- [ ] Brachialis
- [ ] Cleidomastoid
- [ ] Cleidooccipitalis
- [ ] Coracobrachialis
- [ ] Deltoideus
- [ ] Extensor carpi radialis
- [ ] Extensor digitorum communis
- [ ] Extensor digitorum lateralis
- [ ] Flexor carpi radialis
- [ ] Flexor carpi ulnaris
- [ ] Flexor digitorum profundus
- [ ] Flexor digitorum superficialis
- [ ] Latissimus dorsi
- [ ] Omotransversarius
- [ ] Pectoralis profundus
- [ ] Pectoralis superficialis
- [ ] Rhomboideus capitis
- [ ] Rhomboideus cervicis
- [ ] Rhomboideus thoracis
- [ ] Serratus ventralis
- [ ] Serratus dorsalis
- [ ] Supraspinatus
- [ ] Teres major
- [ ] Trapezius
- [ ] Triceps
- [ ] Ulnaris lateralis
- [ ]
- [ ]
- [ ]
- [ ]

Musculature - Pelvic and hindlimb

- [ ] Adductor
- [ ] Biceps femoris
- [ ] Extensor digitorum longus
- [ ] Flexor digitorum longus
- [ ] Gastrocnemius and soleus
- [ ] Gluteus medius
- [ ] Gluteus profundus
- [ ] Gluteus superficialis
- [ ] Gracilis
- [ ] Iliacus
- [ ] Pectineus
- [ ] Peroneus longus
- [ ] Peroneus tertius
- [ ] Psoas major
- [ ] Quadratus femoris
- [ ] Rectus femoris
- [ ] Sartorius
- [ ] Semimembranosus
- [ ] Semitendinosus
- [ ] Tensor fasciae latae
- [ ] Tibialis anterior
- [ ] Tibialis posterior
- [ ] Vastus lateralis
- [ ] Vastus medialis
- [ ]
- [ ]
- [ ]
- [ ]

Additional structures

- [ ]
- [ ]
- [ ]
- [ ]
- [ ]

Fetal Pig - Internal anatomy

Digestive and related systems

- ☐ Anus
- ☐ Cecum
- ☐ Common bile duct
- ☐ Cystic duct
- ☐ Duodenum
- ☐ Epiglottis
- ☐ Esophagus
- ☐ Gall bladder
- ☐ Glottis
- ☐ Hard palate
- ☐ Ileum
- ☐ Jejunum
- ☐ Lesser omentum
- ☐ Liver
- ☐ Mandibular gland
- ☐ Mesentery
- ☐ Nasopharynx
- ☐ Pancreas, left lobe
- ☐ Pancreas, right lobe
- ☐ Parotoid duct
- ☐ Parotoid gland
- ☐ Rectum
- ☐ Rugae of stomach
- ☐ Soft palate
- ☐ Spiral colon
- ☐ Spleen
- ☐ Stomach
- ☐ Sublingual gland
- ☐ Teeth
- ☐ Thymus
- ☐ Thyroid gland
- ☐ Tongue
- ☐
- ☐
- ☐
- ☐
- ☐

Circulatory and respiratory

- ☐ Aortic arch
- ☐ Axillary artery
- ☐ Axillary vein
- ☐ Brachiocephalic trunk
- ☐ Brachiocephalic vein
- ☐ Caudal mesenteric artery
- ☐ Caudal vena cava
- ☐ Celiac artery
- ☐ Cephalic vein, left and right
- ☐ Common carotid artery
- ☐ Coronary artery
- ☐ Coronary vein
- ☐ Costocervical trunk
- ☐ Cranial mesenteric artery
- ☐ Cranial vena cava
- ☐ Deep circumflex artery
- ☐ Deep circumflex iliac artery
- ☐ Deep circumflex iliac vein
- ☐ Deep circumflex vein
- ☐ Deep femoral artery
- ☐ Deep femoral vein
- ☐ Diaphragm
- ☐ Dorsal aorta
- ☐ Ductus arteriosus
- ☐ External iliac artery
- ☐ External iliac vein
- ☐ External jugular veins
- ☐ Femoral artery
- ☐ Femoral vein
- ☐ Gastroduodenal artey
- ☐ Genital artery
- ☐ Genital vein
- ☐ Heart
- ☐ Hepatic artery
- ☐ Internal iliac artery
- ☐ Internal iliac vein
- ☐ Internal jugular veins
- ☐ Larynx
- ☐ Left atrium
- ☐ Left ventricle
- ☐ Lungs
- ☐ Median sacral artery
- ☐ Median scaral vein
- ☐ Mesenteric arteries
- ☐ Mesenteric veins
- ☐ Portal vein
- ☐ Primary bronchi
- ☐ Pulmonary arteries
- ☐ Pulmonary trunk
- ☐ Pulmonary veins
- ☐ Renal artery
- ☐ Renal vein
- ☐ Right atrium
- ☐ Right ventricle
- ☐ Splenic artery
- ☐ Splenic vein
- ☐ Subclavian arteries
- ☐ Subclavian veins
- ☐ Subscapular vein
- ☐ Trachea
- ☐ Umbilical arteries
- ☐ Umbilical vein
- ☐
- ☐
- ☐
- ☐
- ☐

Urogenital systems

- ☐ Bulbourethral glands
- ☐ Cremaster pouch
- ☐ Ductus deferens
- ☐ Epididymis
- ☐ Genital papilla
- ☐ Gubernaculum
- ☐ Horns of the uterus
- ☐ Kidney
- ☐ Medula
- ☐ Ovary
- ☐ Oviduct
- ☐ Penis
- ☐ Prostrate
- ☐ Renal calyces
- ☐ Renal pelvis
- ☐ Renal pyramid
- ☐ Round ligament
- ☐ Scrotum
- ☐ Spermatic cord
- ☐ Suprarenal gland
- ☐ Testis
- ☐ Ureter
- ☐ Urethra
- ☐ Urinary bladder
- ☐ Vagina
- ☐
- ☐
- ☐
- ☐
- ☐

Additional structures

- ☐
- ☐
- ☐
- ☐
- ☐
- ☐
- ☐

Crossword Puzzle - Mammalia

Puzzle solution. An interactive web-based version of this puzzle and its solution are available on the *Digital Zoology* website www.digitalzoology.com/students. With the interactive puzzle you can check to see if individual words or the whole puzzle is correct and get hints for single letters.

Across

1 Mammalian taxonomists spend a lot of time looking at mammal jaws and these when dividing the class into different taxa. (5)

4 These secretions of the sebaceous gland help maintain the hair of mammals. (4)

9 Not all of a mammal's teeth do the same thing, a condition referred to as this. (10)

11 Flying mammals. (4)

13 Describes the soft, dense, insulative hairs. (9)

14 The viviparous placental mammals. (10)

15 The number of sets of teeth that diphyodont mammals have. (3)

17 These are egg-laying mammals. (10)

21 Compared to other vertebrates, this part of the integument is larger and is the biggest part of the mammalian integument. (6)

24 Another term for egg-laying animals. (9)

26 The skin glands that develop at the onset of sexual maturity in mammals. (8)

27 This, along with the intercostal muscles, help aerate the lungs of a mammal. (9)

29 Of the three skull types in vertebrate, mammals have this one. (8)

30 The sharp edges of these mammalian teeth make them ideal for cutting and biting. (8)

31 Hairs in mammals grows out of this. (8)

Down

2 Birds and mammals have to maintain their body temperature at a fixed point with little or no variation. (10)

3 What we're more likely to call the vibrissae, the sensory tactile hairs in mammals. (8)

4 The number of different types of glands that you'll find in mammalian integument. (4)

5 In placental mammals the eggs develop in this part of the female reproductive tract. (6)

6 What an eccrine gland in a mammal secretes. (5)

7 These hairs in mammals protect the fur and provide its coloration. (5)

8 These teeth are important for crushing and grinding food. (6)

10 Metabolism warms the bodies of mammals and birds. This term describes that type of heat. (11)

12 These mammalian glands are used for communication; they're especially large in skunks so that you can be sure to get the message. (5)

16 This type of mammal carries its young in a pouch. (10)

18 The only place will you'll find the ancestral mammalian hairs in a whale. (4)

19 Like feathers, these help insulate mammals and are a diagnostic characteristic of this vertebrate class. (5)

20 Modifications of the apocrine glands to nourish the young are one of the possible origins of these glands in mammals. (7)

22 In mammals the cycle of female fertility. (6)

23 Major protein of mammalian hair. (7)

25 Another term for a mammal's first coat. (6)

28 A mammals first set of teeth are called deciduous, or this type of teeth. (4)

Self-Test - Mammalia

Use the following labels to identify the photographs. You may have to use a label more than once, and some labels may not be appropriate for the photographs. Answers are available on the *Digital Zoology* website at www.digitalzoology.com/students. Be sure to try the interactive drag-and-drop quizzes on the *Digital Zoology* CD-ROM.

Specimen	Labels		
Fetal Pig	1) Ankle	6) Cephalic vein	12) Eye
	2) Aortic arch	7) Cleidomastoid	13) Gall bladder
	3) Auditory canal	8) Cleidooccipitalis	14) Hindlimb
	4) Brachiocephalic vein	9) Common carotid artery	15) Ileum
	5) Brachiocephalicus	10) Ear	16) Internal jugular vein
		11) External jugular vein	17) Left auricle
			18) Left ventricle

19) Liver
20) Masseter
21) Parotid duct
22) Parotid gland
23) Rhomboideus capitis
24) Right auricle
25) Spiral colon
26) Spleen
27) Splenius
28) Sternohyoid
29) Sternomastoid
30) Stomach
31) Subclavian artery
32) Temporalis
33) Thymus gland
34) Thyroid gland
35) Tongue
36) Trunk
37) Umbilical artery
38) Umbilical cord
39) Umbilical vein
40) Wrist

GLOSSARY

4d Cell In animals with determinate cell cleavage, this cell, which appears at the 64 cell stage, forms all of the mesodermal structures.

Abdomen (L. *abdominalis*, pertaining to the abdomen) In organisms that have undergone tagmosis the trunk often develops into two tagma, one involved in locomotion and one that is not locomotory. The tagma not involved in locomotion is referred to as the abdomen.

Abductor (L. *abducere*, to lead away) A muscle that moves any part of the body away from the central midline. In brachiopods, it opens the bivalve shell.

Aboral surface (L. *ab*, away + *os*, mouth) In radially symmetric animals there is no anterior, posterior, left, or right sides to the animal. Instead we refer to the two sides of the animal by the location of the mouth. In this case, the side opposite the mouth.

Abyssal (Gr. *abyssos*, bottomless) The deepest parts of the ocean, usually deeper than 4 kilometers.

Acanthella larva (Gr. *akantha*, spine thorn) The second larval stage in the acanthocephalan life cycle; the acanthella is found inside either an insect or crustacean.

Acanthor larva (Gr. *akantha*, spine thorn) The first larval stage in the acanthocephalan life cycle. It hatches from the egg and uses a ring of hooks around the anterior end to penetrate the first intermediate host, usually an insect or crustacean.

Acanthus (Gr. *akantha*, spine thorn.) The egg and larval stage of an acanthocephalan that passes from the female parasite to the feces of the host.

Acetabulum (L. *acetabulum*, a little vinegar cup) One of two suckers found on the digenic flukes (trematodes). The oral sucker surrounds the mouth, and the acetabulum is located on the ventral surface.

Acoelomate (Gr. *a*, not + *koilos*, hollow) Triploblastic animals that do not have an internal body cavity. This includes the flatworms and ribbon worms. Although the term could be applied to other lower phyla, it is most accurately used with the triploblasts rather than diploblasts.

Acraniate (Gr. *a*, not + *kranion*, skull or head) Chordates that don't have a cranium, or brain case, surrounding the brain. Examples are hagfish and lampreys.

Actinotroch (Gr. *aktis*, ray + *trochos*, wheel) Larval stage of a phoronid. The body consists of a preoral hood formed from the protocoel, a ring of tentacles from the mesocoel, and a metasomal sac from the metacoel.

Actinula A stage in the life cycle of hydrozoans where the sessile polyp is missing. Instead, the planula larva develops directly into an actinula, which then becomes a medusa.

Adaptive radiation Evolution of a variety of different species from a single common ancestor. Each is adapted for a particular niche, and the appearance of the descendants may vary considerably from each other and the ancestor.

Adductor (L. *adducere*, to draw toward) A muscle that moves part of the body toward the central midline. In brachiopods, it closes the bivalve shell.

Advanced characters The traits, or characteristics, that an animal has that are not ancestral to the taxon. They appear later in the evolutionary history of the taxon and cannot be found on the ancestor to the group.

Agamont (Gr. *agamous*, unmarried) In the life cycle of Foraminifera, a diploid agamont undergoes meiosis to produce haploid gamonts. A pair of gamonts will fuse to form a new agamont.

Air sacs In insects, air sacs are part of the respiratory system. Birds have nine air sacs, and their expansion and contraction creates the unidirectional flow of air through the lungs. In whales, air sacs are a part of the nasal passage used to produce sounds used in echolocation.

Akinetic skull (Gr. *a*, not + *kinein*, to move) The type of skull where there is no joint that allows the upper jaw and palate that form the snout or nose to move relative to other parts of the skull.

Allantois (L. *allas*, sausage + *eidos*, shape) One of the extraembryonic membranes found in the amniote animals. The allantois contains the metabolic wastes created by the developing embryo. May also be involved in gas exchange and is important in forming the placenta in placental mammals.

Ambulacral groove (L. *ambulare*, to walk) The groove that runs down the oral surface of each echinoderm arm and contains the tube feet. If the region contains a visible furrow, or groove, they are referred to as open; if not, then closed.

Amebocyte (Gr. *amoibe*, change + *kytos*, cavity or cell) A cell in a multicellular organism that moves using amoeboid movement. Amebocytes often are involved in transporting materials, defending against invading organisms, or cleaning up dead debris and materials.

Amitochondrial (Gr. *a*, not + *mitos*, thread + *chondros*, granule) Without mitochondria. Some protozoa don't have mitochondria, and enzymes for oxidative phosphorylation occur in the cytoplasm.

Ammocoete larva (Gr. *ammos*, sand + *koite*, bed) The larval stage of a lamprey. The ammocoete is of particular interest because it displays the ancestral characteristics of the chordates. During the life cycle of the lamprey, the ammocoete larval stage is found in freshwater streams and later migrates back to the oceans or large lakes.

Ammonia Nitrogenous waste formed from removing the amino group from amino acids. Ammonia is toxic and can only be tolerated in low concentrations. Animals that use ammonia as their metabolic waste also have lots of water available to remove the toxin, which alters pH, reacts with alpha-ketoglutarate in the Krebs cycle to form glutamate, and disrupts the cycle.

Amnion One of the extraembryonic membranes found in terrestrial vertebrate. The amnion is filled with fluid and the developing embryo of reptiles, birds, and mammals develops suspended inside the amnion.

Amniote egg Eggs that shelter the developing embryo in a water-filled sac--the amnion. Characteristic of the amniote animals, which include reptiles, birds, and mammals.

Amphid Unique to the nematodes, these paired sensory structures are located on the sides of the head. An external pore leads to an inner chamber and the sensory receptor formed from modified cilia.

Analogy, Analagous (Gr. *analogia*, proportion) Refers to structures that do not have the same evolutionary origin but have the same functions. The wings of a bat and insect are analogous because both are used for flight. The two structures, however, are not formed from a common structure found in a shared ancestor.

Anapsid (Gr. *a*, not + *apsis*, arch) The type of skull found in reptiles and seen today only in turtles. Behind the opening of the orbits the skull is solid and lacks temporal openings found in other amniote vertebrate skulls.

Ancestral characteristic A character shared by all members of a taxonomic group of organisms (taxon) and used to define the unique nature of the group. The character may be modified or even disappear in some members of the taxon.

Ancestrula (L. *antecedere*, to go before) The founding zooid that undergoes asexual budding to form a bryozoan colony.

Antennae (L. *antenna*, sail yard) Sensory appendages found on the head of a variety of animals, although we commonly associate them with uniramians and crustaceans.

Antennule (L. *antenna*, sail yard) The small sensory biramous appendages found on the first segment of crustaceans.

Apical complex Specialized organelles in parasitic protozoans in the phylum Apicomplexa believed to help the parasite penetrate and enter the host so that they can complete their life cycle.

Apocrine gland (Gr. *krinein*, to separate) A mammalian sweat gland that produces odors and scents used in communication. To release the viscous solution, a part of the secreting cell breaks off.

Appendicular skeleton The bones of the arms, wings, legs, and fins of vertebrates, along with the pelvic and pectoral girdles when present, that attach the limbs to each other and then in turn to the axial skeleton.

Aquiferous system (L. *aqua*, water) This type of system is found in sponges and consists of the canals and chambers through which water flows. Water is pumped through the system by the choanocytes.

Archaeocyte (Gr. *archein*, to begin + *kytos*, cavity or cell) Cell capable of a variety of functions or of differentiating into a variety of cell types with different functions. Examples include the wandering amebocytes in the mesohyl of a sponge.

Archenteron (Gr. *archein*, to begin + *enteron*, intestine) The name given to the primitive gut, the first tube that runs through the developing embryo and is open to the external environment. Formed during gastrulation, it is surrounded by the new endoderm, and it will develop into the digestive system of the organism.

Aristotle's lantern (*Aristoteles*, 384-322 B.C.) The specialized feeding structure is found in the echinoid echinoderms, sea urchins, and sand dollars. Consisting of forty different ossicles, the five teeth are used in feeding.

Arthropodisation (Gr. *arthron*, joint + *podos*, foot) Many of the traits that we consider unique to the Arthropoda may not be independent traits. Instead, they may be the consequence of a tubular exoskeleton. If this is the case, the traits result from a single event, arthropodization--the acquisition of an exoskeleton.

Articulamentum The innermost layer of the shell of a chiton. It has some similarities to the nacreous layer of other molluscs.

Articulata Taxon that includes annelids and arthropods. It is based on the assumption of a natural progression from segmentation seen in annelids to the fusion of segments forming metameres in the arthropods. The taxon Articulata is not supported by the molecular evidence, but is still used in some traditional taxonomic schemes for the animals.

Asconoid Of the three different sponge architectures, this is the simplest. It consists of a central choanocyte lined spongocoel that opens to the outside directly through the osculum. Water enters the spongocoel after passing through the dermal pores.

Asexual (L. *asexualis*, sexless) Organisms that do not reproduce by recombination of genetic material contained in gametes. There is no combination of a haploid sperm and egg to form a zygote.

Asymmetric body plan, Asymmetric A body plan where there is no axis of symmetry that runs through the body and creates identical parts. Examples are sponges and gastropods.

Atriopore (L. *atrium*, entrance hall + Gr. *poros*, pore) The external opening to the atrium. In cephalochordates water passes across the

pharyngeal slits into the atrium and from there leaves through the atriopore.

Auricle (L. *auris*, ear) Chamber of the heart that receives the blood from outside the heart.

Axial skeleton The bones, or cartilage, that make up the skeleton of the main body axis of vertebrates. It includes the cranium, vertebral column, and the rib cages, although not all of these may be present in each of the vertebrate groups.

Axopod (L. *axis*, + Gr. *podus*, foot) Type of pseudopod found on the Actinopod protists. Needlelike, or spinelike, pseudopods are arranged around a central spherical body and supported internally by a microtubular rod, the axoneme.

Benthic (Gr. *benthos*, deep) Animals that live on or close to the bottom of any body of water: lake, pond, stream, river, or ocean.

Bilateral symmetry, Bilaterally symmetric (L. *bis*, twice + *latus*, side) In organisms that have a bilaterally symmetric organization there is only one way that the axis of symmetry can pass through the longitudinal axis and create two identical halves.

Binary fission Cell division where the parent cell divides into two daughter cells equal in size to each other.

Biogenetic law (Gr. *bios*, life +*genetikos*, concerning origin) Ontogeny recapitulates phylogeny. Or, the embryological or developmental sequence of an animal repeats the evolutionary history that gave rise to that animal.

Bioindicator (Gr. *bios*, life) Living organisms that act as indicators of habitat or environmental quality. Many organisms have a narrow range of conditions in which they can live; as environmental conditions change, so does the mix of species that live under these conditions.

Bioluminescence (Gr. *bios*, life + L. *lumen*, light) How living organisms produce light. Many organisms that live in the deep, dark of the oceans, or in other environments at night, use flashing lights, bioluminescence, to communicate.

Bipedal (L. *bis*, twice + *pedis*, foot) Walking upright using the pelvic girdle and limbs.

Biradial symmetry (L. *bis*, twice + *radius*, ray or spoke) The organism appears radially symmetric but at least one set of structures is paired. This results in only two planes of symmetry that pass through the oral-aboral axis of the animal.

Biramous (L. *bis*, twice + *ramous*, branch) A type of appendage in arthropods. Biramous appendages are branched and Y-shaped.

Blastocoel (Gr. *blastos*, germ or shoot + *koilos*, hollow) During the development of the embryo, single cells divide and form a hollow ball of cells one cell layer thick. The cavity inside this hollow ball is the blastocoel.

Blastopore (Gr. *blastos*, germ or shoot + *poros*, opening or pore) The opening to the primitive gut (archenteron) that will develop into either the mouth or anus. The blastopore forms during gastrulation.

Blastula (Gr. *blastos*, germ or shoot) A stage in the development of multicellular animals in which only one cell layer is present, the blastoderm; the cavity the cells surround is the blastocoel. In eggs with little yolk, the embryo resembles a hollow ball; in eggs with yolk, the appearance varies depending on species.

Bony fish The common name for fish in the vertebrate class Osteichthyes. Unlike the cartilaginous fish the skeleton is hardened by the deposition of calcium salts.

Book gill Respiratory structures found in some aquatic cheliciformes. They consist of multiple sheets, or lamellae, across which water flows.

Book lung The respiratory structures found in some spiders. It consists of multiple folds of the cuticle that form sheets, or lamellae, across which air moves for gas exchange.

Branchial basket (Gr. *branchia*, gills) In ascidian adults the pharynx is enlarged to form the branchial basket. Pharyngeal openings in the wall of the basket allow water to pass through with food being trapped on the inner surface before being passed into the gut.

Branchial heart (Gr. *branchia*, gills) Modification of the circulatory system to form secondary hearts that pump blood into the gills.

Branchial vessels (Gr. *branchia*, gills) Branchial describes anything to do with gills, and its combination with vessels describes blood vessels associated with the gills.

Budding A form of asexual reproduction where a small part of the body separates from the parent and develops into a complete organism.

Bursae (L. *bursa*, bag) Paired water-filled sac located at the base of each arm of a brittle star. Cilia in the bursae circulate water. Its surface is important in gas in gas exchange. Gametes may also be released into the bursa.

Byssal threads (Gr. *byssos*, fine flax or linen) In the bivalved mussels these threads help anchor the animal to the substrate, allowing them to live on the surface rather than burrowing into the soft substrate. Byssal threads are made of sclerotized protein.

Captacula (L. *capere*, to capture) The specialized ciliated tentacles that scaphopod molluscs use for feeding. Food is trapped on the sticky bulb at the tip of the captaculum before being moved to the mouth by cilia on the captacula.

Carapace (Sp. *carapacho*, covering or shield) The protective dorsal shield or plate that covers an animal. For example in the Crustacea the cuticle fuses to form a plate over the cephalothorax. In turtles the whole shell is referred to as a carapace.

Carnivore (L. *carni*, flesh + *vorare*, to eat) Any animal that eats other animals as its only source of food.

Caudal (L. *cauda*, tail) A tail or the tail region.

Caudal rami (L. *cauda*, tail + *ramous*, branch) In primitive crustacea the last segment of the body often has paired limblike branches on it. They may be flexible or solid and spinelike.

Cellular grade (L. *cella*, cell or chamber) Organisms with this type of cellular organization are referred to as the Parazoa. They have distinct cells that function independently of each other even though some cells may take on specialized functions. Groups of cells never work together and function as a tissue.

Cephalization Evolution of a distinct anterior region of the body, the head, with specialized sensory structures.

Cephalothorax (Gr. *kephale*, head + *thorax*, chest) In many arthropods the appendages found on the thorax become involved in feeding, and when this happens they are referred to as maxillipeds. The new tagma that results is the cephalothorax.

Cercariae, Cercaria (Gr. *kerkos*, tail) A stage in the life cycle of trematode flukes. The cercaria develops from redia found in the intermediate host. This tadpolelike organism is released from the intermediate host to locate either the primary host or another intermediate host.

Chelicera (Gr. *chele*, claw + *keras*, horn) The unique paired feeding appendage on the first segment of the body of chelicerates such as spiders and ticks.

Chelicifore (Gr. *chele*, claw) These feeding mouthparts are found on most, pycnogonads. Although the proboscis is the main feeding structure in these animals when chelicifores are present, they either hold or tear food and pass it to the proboscis.

Chemosynthetic (Gr. *chemeia*, alchemy + *syntithenai*, to put together) Autotrophic synthesis of complex organic materials based on oxidation of inorganic sulfur compounds rather than light (photosynthesis). Instead of oxygen at the end of the electron transport system, sulfur receives the electron. Common in sulfur bacteria, annelids, and other animals that inhabit the deep sea thermal vent communities.

Chevron Having a V shape.

Chiloped (Gr. *cheilous*, claw + L. *pedos*, foot) The name for a centipedes thoracic leg, to be more precise the leg is a maxilliped, that forms the poison claw used to immobilize their prey. It also gives the taxon its name, Chilopoda.

Chitin (Gr. *chiton*, mantle) A complex carbohydrate composed of linearly arranged N-acetyl-glucosamine units. Chitin is a characteristic of the cell wall of fungi and the outer cuticle of arthropods.

Chitinous (Gr. *chiton*, mantle) A structure composed of chitin, a complex carbohydrate composed of linearly arranged N-acetyl-glucosamine units. Chitin is a characteristic of the cell wall of fungi and the outer cuticle of arthropods

Chloroplast (Gr. *chloros*, light green + *plastos*, molded) Intracellular organelle in photosynthetic plants and protists (algae) that captures light energy using photosynthesis.

Choanocyte (Gr. *choane*, funnel + *kytos*, cavity or cell) These unique collar-shaped cells whose flagella are responsible for generating the water current in the sponge. As the flagella beat, food particles are trapped against the microvilli that form the collar. Choanocytes are also found in some colonial protists.

Choanoderm (Gr. *choane*, funnel + *derma*, skin) The layer of choanocyte cells lining the different parts of the aquiferous system in a sponge: in the asconoid architecture, the spongocoel; in syconoid, the radial canals; and in leuconoid sponges, the choanocyte chambers.

Choanoflagellate (Gr. *choane*, funnel + L. *flagellum*, whip) The protozoans that resemble the choanocyte cells found in sponges. They have a collar of microvilli that surround a central flagellum. When the flagellum beats, food is trapped against the collar and consumed by phagocytosis.

Choridoid larva In the recently discovered Cycliophora, the choridoid larva is part of the sexual cycle. After mating, this larval stage escapes and settles on a new host. It is how the cycliophorans get back onto their lobster host after the host molts.

Chorion (Gr. *chorion*, membrane) The outermost extraembryonic membranes in reptiles, birds, and mammals. It is involved in gas exchange. In insects, the chorion is the outer shell of the egg secreted by the follicle cells of the ovary.

Cilia (L. *cilium*, eyelash or eyelid) Cellular hairlike locomotory structures that consist of an extension of the plasma membrane surrounding a 9+2 organization of microtubules. Unlike flagella, cilia are shorter and more numerous on the cell surface.

Cirri (L. *cirri*, curl) Slender filamentous structures: fused cilia in protozoa, attachment appendages in sessile echinoderms, or a copulatory organ found in some invertebrates.

Clade (Gr. *klados*, branch sprout) In cladistic taxonomy, how groups of animals are related to each other. It includes the ancestral group and all of the other groups, or lineages, of animals that arose from the ancestral group. Clades are monophyletic.

Cladistics (Gr. *klados*, branch sprout) A method for classifying organisms based on primitive and derived characters. The resulting arrangement of organisms reflect evolutionary relationships between the taxa.

Cladogram (Gr. *klados*, branch sprout + *gramma*, mark or drawing) A visual representation of the phylogenetic branching of different animal groups based on cladistics. There are no units or measures to the dimensions of a cladogram.

Clitellum (L. *clitellae*, back or saddle) Special fused metameres in oligocheates and leeches that secrete mucus during mating.

Cloaca (L. *cloaca*, sewer) The shared common opening to a number of organ systems that may include the reproductive, digestive, and/or excretory system.

Cloacal siphon (L. *cloaca*, sewer + Gr. *siphon*, tube, pipe) Part of the ascidian body that surrounds the point where the water leaves the animal--excurrent flow. Water is pumped the action of the cilia lining the pharyngeal basket.

Closed circulatory system A circulatory system in which the circulating fluid (blood) flows in vessels or tubes connected to each other by capillaries.

Cnidae (Gr. *knide*, nettle) Describes the different types of cnidocytes found in the cnidarians.

Cnidocil (Gr. *knide*, nettle + L. *cilium*, eyelash or eyelid) A modified flagellum on the cnidocyte that causes the nematocyst inside to fire. The stimulus involves some sort of chemical cue; touching the cnidocil doesn't fire the nematocyst.

Cnidocyte (Gr. *knide*, nettle + *kytos*, cavity or cell) Specialized cells found only in the Cnidaria. When these cells evert, a nematocyst is discharged. The nematocyst may act as a stinger or a sticky thread to entangle and capture prey.

Cochlea (Gr. *kochlos*, snail with a spiral shell) Bone in the inner ear that converts vibrations from the eardrum into nervous impulses interpreted as sounds. Found in mammals and birds.

Coelenteron (Gr. *koilos*, hollow + *enteron*, intestine) The name given to the internal cavity of the cnidarians. This is an incomplete gut with only one opening, the mouth. Food to be digested and undigested food that must be eliminated both pass through the mouth. The cavity is lined by gastrodermis.

Coelom (Gr. *koiloma*, cavity) A true body cavity completely lined by mesoderm, which forms the peritoneum. Animals with true coeloms are referred to as eucolomates.

Coelomate (Gr. *koiloma*, cavity) Animals that have a true coelom entirely lined with mesoderm. Coeloms are formed in two ways: enterocoelic pouches or by splitting a block of mesoderm, schizocoely, to create the coelom inside.

Coelomocytes (Gr. *koiloma*, cavity + *kytos*, cavity or cell) Cells found in coelomic fluid. They are often undifferentiated amebocytes.

Coelomoduct (Gr. *koiloma*, cavity) Tubules that connect the coelomic cavity with the outside of the animal.

Coenenchyme In some colonial corals the polyps sit on a mass of mesoglea, the coenenchyme, covered in pinacoderm or epidermis. Extensions of the gastrovascular cavities from different polyps in the colony weave through the coenenchyme and connect with each other.

Coenosarc (Gr. *koinos*, common + *sarx*, meat) In colonial hydrozoan cnidarians, polyps are connected to each other by hollow tubes, an extension of the gastrovascular cavity of each polyp in the colony. The coenosarc is the living part of the connections between polyps.

Collagen (Gr. *kolla*, glue + *gennan*, to produce) Tough fibrous protein found in the connective tissue of vertebrates and the cuticles of some invertebrates. The protein is flexible but doesn't stretch or compress.

Collagenous (Gr. *kolla*, glue + *gennan*, to produce) Made from collagen, a fibrous protein with high concentrations of glycine, proline, alanine, and hydroxyproline. If you boil collagen, it becomes gelatin.

Colloblast cell (Gr. *kolla*, glue + *blastos*, bud or germ) The unique sticky discharge cells found in the tentacles of ctenophorans. They are used to capture prey, and although cnidoblasts resemble cnidocytes, this is a convergent trait making the two analogous structures.

Collum Unique structure in millipedes. It is formed from the first trunk segment, which has lost its appendages and become enlarged into a collarlike segment that surrounds the head.

Columella (L. *columna*, column or pillar) The central axis in the whorl of a gastropod (snail) shell.

Comb rows Fused cilia used by ctenophorans for locomotion. The comb rows are arranged in bands connected to the apical organ that coordinates the beat of the comb rows.

Commensal (L. *cum*, with + *mensa*, table) A kind of symbiosis where one animal lives on or in another and benefits from living there. This type of relationship has no effect on the host, which is neither at a disadvantage nor an advantage because of the commensal organism.

Compensation sacs Responsible for the hydrostatic inversion of the feeding introvert.

Complete alimentary tract, Complete digestive system, Complete digestive tract, Complete gut (L. *dis-*, apart + *gerere*, to carry) A digestive system that has an anal and oral opening. This adaptation allowed for a linear processing of ingested food with specialized regions in the gut for grinding, mixing, and digesting food under different conditions.

Complete metamorphosis (Gr. *morphosis*, to form or make) Found in insects where the life cycle includes larval, pupal, and adult stages. The larval stage looks nothing like the adult, and during the pupal stage there is a complete reorganization of tissues and structures inside to create the transformation from larva to adult. Examples: maggots and flies, caterpillars and butterflies, grubs and beetles.

Complex ciliature (L. *cilium*, eyelash or eyelid) Specialized ciliary structures found in the higher ciliates (Ciliophora). The single 9+2 cilia fuse with others to form larger more complex structures such as cirri and membranelles.

Compound eye The characteristic eye of the Arthropoda. It consists of many ommatidia, the basic optical unit of the compound eye, grouped together to form the compound eye. With the exception of crustaceans, the compound eyes are usually found in combination with much simpler single lens ocelli.

Concurrent exchange When two fluids flow in the same direction as something, heat or oxygen for example, diffuses between them. Although there may be a concentration difference at the start, equilibrium is reached. Once this occurs, diffusion stops even though the two fluids still flow past each other.

Conjugation (L. *con*, together + *jugare*, to join) A form of sexual reproduction in ciliates (Ciliophora). During conjugation two ciliate protozoans join and the macronuclei disappear. After meiotic divisions of the micronucleus, the resulting genetic material is exchanged.

Convergent evolution (L. *convergere*, to incline together) When two distinct evolutionary lineages do not share a common ancestor but evolve similar structures or features. These features are considered analogous.

Convergent trait (L. *convergere*, to incline together) A trait or characteristic similar in two animals from two individual evolutionary lineages. These characters are considered analogous.

Coracidium (Gr. *korax*, raven) Ciliated larval stage of tapeworms that have an aquatic stage in their life cycle. The coracicium invades a crustacean copepod and becomes a procercoid larva.

Corona (Gr. *korone*, crown, circle) The crown formed from a ciliated disc at the anterior end of a rotifera.

Countercurrent exchange When two fluids flow in opposite directions as something, heat or oxygen for example, diffuses between them. Because the concentration difference between the two solutions never reaches equilibrium, diffusion continues along the entire surface where the fluids pass by each other.

Coxal gland (L. *coxa*, hip) The excretory gland found in some spiders and noninsect uniramians. This blind-ended sac is bathed in hemolymph, and wastes are excreted at the base of the coxal segment of the leg.

Craniate (Gr. *kranion*, skull) Vertebrate animals that have a brain case of either cartilage or bone.

Cranium (Gr *kranion*, skull The brain case, made of either cartilage or bone, that surrounds the brain.

Cristae (L. *crista*, crest, comb, or tuft) The folds of the inner membrane of the mitochondria. The protein complexes of the electron transport chain are located on the cristae.

Crop Any portion of the anterior part of the digestive tract used to store food prior to its digestion.

Cryptobiosis (Gr. *kryptos*, hidden + *bios*, life) A dormant, or suspended, state of animation that allows some animals to survive severe conditions such as extremely low temperatures. Dehydration occurs in preparation for the suspended state and rehydration is associated with the return of favorable conditions.

Crystalline style (Gr. *krystallos*, ice or crystals) A rodlike structure in some mollusc stomachs made of enzymatic proteins required for digestion. Cilia lining the stomach rotate the crystalline style, which grinds against the gastric shield to release the digestive enzymes.

Ctene (Gr. *kteis*, comb) This unique locomotory structure is found in the Ctenophora and is made up of fused cilia arranged into flattened plates. The ctenes are then organized into eight bands that run between the oral to aboral surface of the animal.

Ctenidia (Gr. *kteis*, comb + *-idion*, dim) Molluscan gills that often have additional functions other than respiration.

Ctenoid (Gr. *kteis*, comb + *eidos*, form) Bony, dermal scales found in the advanced fishes (teleosts). The anterior edge of the scale is embedded in the skin. The posterior edge is free with a tooth- or comblike edge that may be important in reducing drag as water flows over the surface of the fish.

Cuticle (L. *cutis*, skin) The nonliving and noncellular outer layer of an organism secreted by the underlying epidermis. Cuticles are common in a variety of animals including nematodes, annelids, and arthropods. The presence of a cuticle precludes the presence of cilia.

Cuvierian tubules (Georges Cuvier, 1769-1832) Defensive organs used by some sea cucumbers. They are located at the base of the respiratory tree and are shot out of the cloaca if the animal is threatened.

Cycloid (Gr. *kyklos*, circle) Bony, dermal scales found in advanced fishes (teleosts). The scales are arranged in rows. The anterior edge of the scale is embedded in the skin, and the free posterior edge is smooth and round.

Cydippid (*Kydippe*, a girl in Greek mythology) The larval stage of free-spawning ctenophorans. Cydippids usually have spherical symmetry and bear a superficial resemblance to the adult.

Cystacanth (Gr. *kystis*, pouch, sac, bladder, or cyst + *akantha*, spine thorn) A stage in the life cycle of an acanthocephalan. After the acanthella forms in the intermediate hosts, it encysts as the cystacanth that infects the definitive host when it is consumed.

Cysticercus (Gr. *kysitis*, pouch, sac, bladder, or cyst + *kerkos*, tail) Larval stage in tapeworms without an aquatic stage in the life cycle. The csyticercus is usually found in the muscle of the intermediate mammalian host and is ingested by the definitive host. It consists of an inverted scolex inside the bladderlike cyst.

Cystid In bryozoan (ectoproct) colonies the cystid consists of the nonliving shell, or casing, and the living part of the animal that secretes the casing. The cystid doesn't include the lophophore and the part of the body that moves inside the cystid.

Cytostome (Gr. *kytos*, cavity or cell + *stoma*, mouth) In some protozoans, especially ciliates, phagocytosis always occurs at the same position on the cell surface. When this occurs the location is referred to as the cytostome (cell mouth).

Dactylozooid (Gr. *daktylos*, finger or toe + *zoon*, animal + *eidos*, form) Polyps in colonial hydrozoans specialized for defense using batteries of stinging cnidocytes.

Definitive host In parasites with complex life cycles that involve more than one host, the organism that contains the adult stage of the parasite.

Dendrogram (Gr. *dendron*, tree + *graphein*, to write or record) A visual representation of evolution that branches like a tree to depict the relationships between the different groups.

Dentary (L. *dens*, tooth, plural *dentes*, teeth) Jaw bone of mammals formed from the fusion of smaller bones. In other vertebrates, the dentary is one of a number of bones that form the lower jaw.

Derived characteristic A characteristic modified or altered from the original ancestral form.

Dermal branchiae (Gr. *derma*, skin + *branchia*, gills) External extensions of the outer epidermis and peritoneum of the echinoderm body cavity. Both the outer epidermis and inner peritoneum are lined with cilia. The surface of the dermal branchiae are important in diffusion of gases and metabolic wastes.

Dermis (Gr. *derma*, skin) The inner layer of the epidermis consisting of connective tissues; vascular tissue, blood, and lymphatic vessels; and sensory cells.

Detorsion In gastropod snails torsion involves a 180-degree rotation of the visceral mass relative to the underlying food. Some snails reverse the rotation--detorsion.

Deuterocerebrum (Gr. *deuteros*, second + L. *cerebrum*, brain) When present, it's the second, or middle, part of the arthropod brain. It innervates the first pair of antenna in crustaceans, the antenna in uniramians, and is missing in chelicerates that don't have antennae. A chelicerate brain is the fusion of only two segments.

Deuterostome (Gr. *deuteros*, second + *stoma*, mouth) Phyla, including the Chordata and Echinodermata, that share common characteristics of the blastopore not forming the mouth, radial indeterminate cellular cleavage in the embryo, and the formation of the body cavity by enterocoelic pouching.

Diapsid (Gr. *dis*, twice + *apsis*, arch) Vertebrate skull of some reptiles and all birds. There are two temporal openings behind the orbits in the skull.

Digestive cecum (L. *dis-*, apart + *gerere*, to carry + L. *caecum*) A blind-ended pouch that extends from the main digestive tract. Digestive ceca may be the sites for final digestion of ingested food or may be regions of the gut with specialized enzymes or conditions required for digestion.

Digestive gland (L. *dis-*, apart + gerere) Many invertebrates, including molluscs, arthropods, and echinoderms, have pockets or evaginations from the main alimentary tract where specialized digestive events occur.

Dimorphic life cycle (Gr. *dis*, two + *morphe*, shape) When the life cycle of the animal includes two distinct and physically different body types. The life cycle progresses from the immature form to the reproductive type. Or both forms may appear together in colonial types of dimorphic organisms.

Dimorphic nuclei (Gr. *dis*, two + *morphe*, shape) Organization of the nuclear material typical of ciliates. There is always a single small micronucleus and either one large or many macronuclei. Micronuclei are involved in reproduction and the macronuclei are used by the cell to function.

Dimorphism (Gr. *dis*, two + *morphe*, shape) When the life cycle of the animal includes two distinct and physically different body types. The life cycle progresses from the immature form to the reproductive type. Or both forms may appear together in colonial types of dimorphic organisms.

Dioecious (Gr. *dis*, two + *oikos*, house) Organisms that have the male or female reproductive structures in separate individuals. This is the opposite of monoecious.

Diphyodont teeth (Gr. *diphyes*, twofold + *odous*, tooth) The type of teeth found in mammalian vertebrates with two sets of teeth--a baby set replaced with permanent adult teeth.

Diploblastic (Gr. *Diplos*, double + *blastos*, germ or shoot) Organisms formed from only the two primitive cell layers--endoderm and ectoderm. Although there may be some type of a matrix between the two cell layers, often referred to as mesoglea or mesenchyme, it is not a true tissue layer.

Diploid (Gr. *Diploos*, double + *eidos*, form) Cell or stage in the development of the organism, where the cell nucleus contains two pairs of homologous (2N) chromosomes.

Diplosegment (Gr. *Diploos*, double) A characteristic of the millipedes, Diplopoda, where adjacent segments have fused in pairs. As a result there appears to be two pairs of appendages for each of the visible segments of the trunk.

Direct development Type of development when a miniature adult hatches from the egg instead of a free, often swimming, larval stage that grows and undergoes a transformation into the adult form.

Dissolved organic matter Organic material dissolved in water and not in a particulate form.

Diverticulum (L. *divertere*, to turn aside) Any hollow branch or pouch that extends from the side of an organ.

Dorsal (L. *dorsum*, back) In animals with bilateral symmetry, the upper surface of the animal if it crawls across the substrate and the back of the animal if it walks upright.

Dual circulatory system In vertebrates such as amphibians, reptiles, birds, and mammals, blood flows through two different paths or circuits. One to the lungs, pulmonary, and the other to the rest of the body, systemic.

Dual gland adhesive system Glandular system found in some invertebrates, notably Platyhelminthes. One gland secretes glue that holds the animal in place; a different gland secretes a second compound to dissolve the glue and break the attachment.

Eccrine gland (Gr. *ek*, out + *krinein*, to separate) Glands that secrete materials using exocytosis without destroying the gland cell. An example is the sweat glands, which produce a watery secretion in mammals.

Ecdysial gland (Gr. *ekdysis*, getting out or escape) Gland secreting the hormone ecdysone, which controls molting in arthropods.

Ecdysis (Gr. *ekdysis*, getting out or escape) The periodic molting, or shedding, of the outer exoskeleton of an arthropod.

Ecdysozoa Taxon of animals that have a chitinous or collagenous cuticle, molt, and no epithelia cilia. The Ecdysozoa is a new taxon identified by molecular techniques, and it breaks the link between the Annelida and Arthropoda (Articulata) used in traditional evolutionary classification of animals.

Ectoderm (Gr. *ektos*, outside + *derma*, skin) The outermost cell layer that forms the epithelium and nervous systems of an animal. It, and the endoderm, are the two primary germ layers.

Ectoparasite, Ectoparasitism (Gr. *ektos*, outside + *parasitos*, parasite) Parasitic organisms depend on another animal, their host, for survival. Ectoparasites live on the outer surface of the host. This is the opposite of an endoparasite, which lives inside the host.

Ectoplasm (Gr. *ektos*, outside + *plasma*, form) The outer cytoplasm of a cell or protozoan that has no organelles and is usually less fluid than the inner endoplasm.

Ectotherm (Gr. *ektos*, outside + *therme*, heat) Organisms unable to generate body heat internally using metabolic processes and instead use their environment. The opposite is endothermic.

Elastic capsule chromatophore (Gr. *chroma*, color + *phorein*, to carry) Specialized pigmented cells on the surface of cephalopods that, by changing their shape, expose differing amounts of pigment changing the color and appearance of the cephalopod.

Electron transport chain Series of protein complexes in mitochondria, or chloroplasts, which pass electrons between the complexes. It involves a series of oxidation and reduction reactions that pass the electron to oxygen to form water and ATP.

Endocuticle (Gr. *endon*, within + L. *cuticula*, thin skin) Innermost layer of the arthropod procuticle consisting of proteins and chitin. Endocuticle is not hardened, or sclerotized, and there are no covalent interactions between the proteins and chitin. Endocuticle is digested and reused to build new procuticle during the molt.

Endoderm (Gr. *endon*, within + *derma*, skin) The innermost layer of cells that forms the digestive tract and other associated organs. The ectoderm and the endoderm form the two primary germ layers of an animal.

Endoparasite (Gr. *endon*, within + *parasitos*, parasite) Parasitic organisms that depend on their host for all, or at least one stage, of their life cycle. Endoparasites complete this requirement inside the host.

Endoplasm (Gr. *endon*, within + *plasma*, form) The inner cytoplasm of a cell or protozoan with organelles; usually more fluid than the outer ectoplasm.

Endopodite (Gr. *endon*, within + *podos*, foot) The inner branch of the biramous appendage in crustaceans.

Endoskeleton (Gr. *endon*, within + *skeletos*, hard) Supporting structures of the skeleton surrounded by the body tissues. As a consequence, there is living tissue on all sides of endoskeletal structures. Endoskeltons are found, for example, in echinoderms and chordates.

Endostyle (Gr. *endon*, within + *stylos*, pillar) A ciliated groove on the ventral surface of the pharynx in early chordates and related taxa. Mucus produced by the endostyle traps particulate food, and the cilia propel it into the digestive tract.

Endotherm, Endothermic (Gr. *endon*, within + *therme*, heat) Organisms able to generate body heat internally using metabolic processes. The opposite is ectotherm.

Ephyra (Gr. *ephyra*, sea nymph) Free-swimming, miniature medusa of scyphozoans produced by asexual budding from the scyphistome.

Epicuticle (Gr. *epi*, on or upon + L. *cuticula*, thin skin) The outermost part of the arthropod cuticle that lacks any chitin. The epicuticle forms a chemical barrier, and the waxes in it prevent cuticular water loss in terrestrial arthropods.

Epistome (Gr. *epi*, on or upon + *stoma*, mouth) Flaplike covering of the oral opening in lophophorates. Formed from the protocoel, it may or may not have a coleomic cavity inside.

Epitheliomuscular cell Cells that line the outer surface of cnidarians. These cell have two functions: the first is to form the outer body covering of the animal, the second is in movement by contraction of the myoneme portion of the cell.

Epithelium (Gr. *epi*, on or upon + *thele*, nipple or thin skin) Layer of cells covering the outer surface of an organism or lining the free surface of internal vessels, ducts, and cavities.

Epitoke (Gr. *epitokos*, fruitful) Marine polychaete worms that, during the breeding season, develop gonads and become filled with gametes. This may involve epitoke segments or whole epitoke worms.

Errant Mobile and actively moving organisms.

Esophagus (Gr. *oissein*, to carry + *phagein*, to eat) Part of the digestive system that carries food from the pharynx to the stomach.

Eucoelomate (Gr. *eu*, good or true + *koilos*, hollow) (Coelomate) Organisms that possess a true coelom with mesoderm lining the whole body cavity, unlike pseudocoelomates where the mesoderm is adjacent only to ectodermal tissue.

Eukaryote (Gr. *eu*, good or true + *karyon*, nucleus or nut) Cells, or organisms with cells, containing a membrane-bound nucleus.

Eutely (Gr. *euteia*, thrift) Also referred to as cellular constancy. Eutelic animals are often composed of a fixed number of cells, or nuclei, in the adult, and this number is species specific. Eutely is often associated with animals that have specialized for miniaturization.

Evisceration A presumed defensive strategy used by sea cucumber. When threatened, they shoot parts of the digestive tract and other organ systems out the cloaca at their attacker.

Evolutionary taxonomy Based primarily on the principles of evolution. New taxa result from genetic isolation or niche specialization.

Exocuticle (Gr. *exo*, out or outside + L. *cuticula*, thin skin) Outermost layer of the arthropod procuticle consisting of proteins and chitin. Exocuticle is hardened, or sclerotized, with phenolic compounds that chemically link the proteins in the endocuticle. Exocuticle is resistant to digestion and can't be reused during the molt.

Exopodite (Gr. *exo*, out or outside + *podos*, foot) The outer branch of the biramous appendage in crustaceans.

Exoskeleton (Gr. *exo*, out or outside + *skeletos*, hard) Supporting structures of the skeleton not surrounded by the body. Endoskeletal elements are secreted by an underlying epidermis and one side of the skeletal structures is exposed outside the body. Found, for example, in arthropods.

Extant (L. *exstare*, to stand forth) Species, or groups of animals, that are still living today; they aren't extinct.

Extinct (L. *exstinguere*, to put out or extinguish) Animals that existed in the past but are no longer living species.

Extracellular digestion (L. *extra*, outside + *cell*, compartment or cell) The final breakdown of ingested food to components small enough to be absorbed across a cell membrane occurs outside of the cell in the lumen of the digestive tract--intestine or digestive gland, for example.

Extracorporeal (L. *extra*, outside + *corpora*, body) Outside of the body. An example is extracorporeal digestion found in some animals where digestive enzymes are regurgitated into the food. Once it has been broken down, the resulting product is consumed.

Extraembryonic membranes (L. *extra*, outside + Gr. *embryon*, embryo) A set of membranes that surround the developing embryo in reptiles, birds, and mammals. They include the amnion, chorion, allantois, and yolk sac.

Extrusome Cellular organelle discharged from the cell surface of a ciliate; trichocysts, for example.

Eye cup Simple eye in invertebrates, eye spot, surrounded by pigment cells that give it a cup-shaped appearance.

Eye spot Simple eye found in invertebrates; consisting of photoreceptive cells positioned underneath a lens that can either change its shape or be positioned to focus light. May be image-forming on some (nautilus and spiders), or it may only detect light intensities.

Filopod (L. *filum*, thread + Gr. *podos*, foot) Type of pseudopod consisting of thin ectoplasm strands without any internal support and doesn't form a net or mesh with other pseudopods.

Finiculus (L. *funis*, cord) A cord connecting the digestive system and wall of a zooid in some bryozoan colonies. It may be important in exchange of materials between individuals in the colony.

Flagella A cellular hairlike locomotory structure that consists of an extension of the plasma membrane surrounding a 9+2 organization of microtubules. Unlike cilia, flagella are longer and are usually found singly or in pairs on each cell.

Flame cell Alternate name for the protonephridia based on their appearance as flickering flames. The flicker is from the cilia that beat inside the cap cell of the protonephridium.

Forelimb The appendages attached to the anterior pectoral girdle of vertebrates. The forelimbs may be modified as legs, arms, wings, or fins.

Gall bladder Part of the vertebrate digestive system; an enlargement of the bile duct secreting bile and bile salts into the small intestine to emulsify fats facilitating the digestion of the fat.

Gamete (Gr. *gamos*, marriage) One of two types of sex cells, egg or sperm, that fuse to form the zygote.

Gamont (Gr. *gamos*, marriage) A protozoan gamete in protists with sexual reproduction.

Ganglion (Gr. *ganglion*, knot) A mass of nerve cells that forms an enlarged knotlike structure.

Ganoid (Gr. *ganos*, shinny + *eidos*, form) Thick, square-, diamond-, or rhombiod-haped fish scales that don't overlap. Found in primitive fishes.

Gastrodermis (Gr. *gaster*, belly or stomach + *derma*, skin) The name given to the endodermal cells that line the gastrovascular cavity (coelenteron) of cnidarians.

Gastrovascular cavity (Gr. *gaster*, belly or stomach + L. *vasculum*, small vessel) The name given to the internal cavity of the cnidarians. A blind-ended (incomplete) gut with only one opening. Food to be digested and undigested food that must be eliminated pass through the mouth. The cavity is lined by gastrodermis.

Gastrozooid (Gr. *gaster*, belly or stomach + *zoon*, animal + *eidos*, form) Polyps in colonial hydrozoans specialized for feeding; also referred to as hydranths.

Gastrula (Gr. *gaster*, belly or stomach) Stage in embryonic development where the embryo consists of two cell layers, an outer ectoderm and an inner endoderm, surrounding the primitive gut (archenteron) opening to the outside through the blastopore.

Gastrulation (Gr. *gaster*, belly or stomach) During embryological development this stage results in the blastulas conversion into a gastrula. Cells migrate toward the inside of the embryo from the region where the blastopore will form to create the second germ layer (endoderm). The embryo changes from having only one cell layer to having two cell layers.

Gemmule (L. *gemma*, bud) A bud of asexually formed freshwater sponge cells resistant to unfavorable conditions; permits the organism to survive adverse conditions.

Gill arches Supports for the gills in chordates. They form from the tissues between the pharyngeal slits and may include additional support from cartilage or bone.

Gill bar Supports for the gills in chordates. They form from the tissues between the pharyngeal slits and may be strengthened with cartilage or bone.

Gill raker Bony structures in the inner gill surface of jawed fishes that keep ingested food inside the buccal cavity and pharynx.

Gill slit Lateral opening in the wall of the pharynx that allows water to enter into the mouth and exit through the pharynx. This is one of the ancestral characteristics of the phylum Chordata. It allowed water to be removed from ingested food before it was passed back into the digestive system.

Gizzard (L. *gigerium*) Muscular, grinding part of the digestive system. In birds, it is an enlargement of the anterior part of the stomach; in invertebrates, it is part of the foregut that may also be lined with cuticular plates and teeth.

Glochidium (Gr. *glochis*, arrow point) Free-swimming, larval stage of some freshwater, bivalve molluscs. Glochidia are brooded inside the gill chamber. After release, they parasitize fish until the glochidia's development is complete.

Glomerulus (L. *glomus*, ball) Part of the vertebrate kidney consisting of capillaries surrounded by the renal capsule; also a network of blood sinuses forming a spongy tissue in hemichordates and assumed to have an excretory function.

Gnathobase (Gr. *gnatho*, jaw) In many arthropods the base of the appendages is modified into grinding surfaces used to break up food before it is passed to the mouth.

Gnathochilarium (Gr. *gnatho*, jaw + *cheilos*, lip) Structure formed by the fusion of the first maxillae. This flaplike structure is found in millipedes, diplopods, and pauropauds.

Gonad (Gr. *gone*, seed) The reproductive organ that produces gametes, eggs or sperm, by meiotic reduction of the chromosome content.

Gonoduct (Gr. *gone*, seed) Duct or tube that carries gametes from the gonads to the outside of the animal.

Gonophore (Gr. *gone*, seed + *phorien*, to carry along) Sexually reproductive, medusal bud that develops in some colonial hydrozoans. The resulting mature medusa may be released from the colony or remain attached when they release their gametes.

Gonopore (Gr. *gone*, seed + *poros*, opening) External opening to the reproductive tract.

Gonozooid (Gr. *gone*, seed + *zoon*, animal + *eidos*, form) Polyp in colonial hydrozoans specialized for producing medusa, the reproductive stage in the life cycle. Gonozooids are also referred to as gonangia.

Gravid (L. *gravis*, heavy) An organism either filled with eggs or pregnant.

Haploid (Gr. *haploos*, one fold or single) Cell, or stage in the development of the organism, where the cell nucleus contains only one (N) of the pairs of homologous chromosomes.

Haptor (Gr. *haptein*, to fasten) Holdfast that monogenean flukes use to hold onto their hosts; includes a prohaptor and opisthaptor.

Hemal system (Gr. *haima*, blood) System of vessels and spaces in echinoderms; function is uncertain. In sea cucumbers, it is well developed and connects the respiratory tree and intestine suggesting a possible circulatory role.

Hemerythrin (Gr. *haima*, blood + *erythros*, red) A violet-colored respiratory pigment that uses iron in the oxygen-binding site. Hemerythrins are most often used to store oxygen.

Hemimetabolous (Gr. *hemi*, half + *metabole*, change) Insects that lack a pupal stage in the life cycle.

Adults resemble the larval stages in appearance but have wings and a functional reproductive system.

Hemocoel (Gr. *haima*, blood + *koilos*, cavity) The principle body cavity in molluscs and arthropods, remnant of the blastocoel. It forms part of the open circulatory system found in these animals. The true coelom is usually reduced to a cavity surrounding the heart, the pericardial cavity, or gonads.

Hemocyanin (Gr. *haima*, blood + *kyanos*, blue) Blue-colored respiratory pigment that uses copper to bind oxygen; found in invertebrates, including molluscs and some crustaceans.

Hemoglobin (Gr. *haima*, blood + L. *globus*, globule) Respiratory pigment that binds oxygen using iron molecules in a heme structure; common in vertebrates and some invertebrates, such as the terrestrial worms.

Hemolymph (Gr. *haima*, blood + *lympha*, water) The fluid contained in the hemocoel of organisms with an open circulatory system. Hemolymph acts as both the coelomic and circulatory fluids.

Herbivore (L. *herba*, grass or herb + *vorare*, to devour) Any animal that eats plants as its only source of food.

Heterocercal (Gr. *hetero*, other or different + *kerkos*, tail) The type of tail fin in a fish where the dorsal lobe is larger than the ventral lobe and the vertebrae of the axial skeleton form part of the dorsal lobe. This type of tail is typical of sharks.

Heterodont (Gr. *hetero*, other or different + *odous*, teeth) When a jaw has different types of teeth, each specialized for a different function.

Heterotherm (Gr. *hetero*, other or different + *therme*, heat) Animals whose body temperature fluctuates. Fluctuations in body temperature are often related to temperature changes in the surrounding environment.

Heterotroph, Heterotrophic (Gr. *hetero*, other or different + *trophe*, nutrient, nourishment) Organisms not capable of converting light into chemical energy. They must consume other organisms or material produced by other organisms to survive. This is a characteristic of all animals.

Hexaradially symmetric (Gr. *hex*, six) Symmetry based on six; found in some corals.

Hexaxon (Gr. *hex*, six) Type of sponge spicule made of silica and with six points; characteristic of the glass sponges (Hexactinellida).

Higgins' larva Feeding larval stage of a loriciferan.

Hindlimb The appendages attached to the posterior pelvic girdle of vertebrates. The hindlimbs may be modified as legs or fins.

Holometabolous (Gr. *holos*, whole + *metabole*, change) Insects that have a pupal stage in the life cycle. Adults don't resemble the larval stages in appearance. The pupal stage, between the larval and reproductive adult, allows for the complete reorganization of the body plan.

Homeotherm (Gr. *homos*, same + *therme*, heat) Organisms that maintain their body temperature at stable temperatures independent of the surrounding environmental temperature.

Homology, Homologous (Gr. *homologia*, agreeing or corresponding) Structures that have a similar evolutionary origin but different functions. The wing of a bat and a whale's flipper are homologous because both are derived from the anterior appendages of the ancestral vertebrate--one is used for flying, the other for swimming.

Homonomous (Gr. *homos*, same + *nomos*, law) Series of repeated structures identical in appearance.

Hydromedusa (Gr. *hydro*, water + *Medusa*, creature in Greek methology) The medusa in a hydrozoan.

Hydrorhiza (Gr. *hydro*, water + *rhiza*, root) The rootlike base of a hydrozoan colony that attaches the colony to the substrate.

Hydrostatic skeleton (Gr. *hydro*, water + *statikos*, standing) Formed from a fluid-filled and closed cavity surrounded by a body wall containing muscles oriented in different directions. Muscular contractions maintain the rigid form or change the shape of the organisms allowing movement.

Hyperosmotic (Gr. *hyper*, above + *osmos*, push) When the salinity, osmotic pressure, of two solutions is compared, the one with the higher salinity is hyperosmotic to the other.

Hypertonic (Gr. *hyper*, above + *tonos*, tension) When the salinity, osmotic pressure, of two solutions is compared, the one with the higher salinity is hyperosmotic to the other and referred to as being hypertonic.

Hypoosmotic (Gr. *hypo*, under + *osmos*, push) When the salinity, osmotic pressure, of two solutions is compared, the one with the lower salinity is hypoosmotic to the other.

Hypostracum (Gr. *hypo*, under + *ostrakon*, shell) An alternate term for the inner nacreous layer of a mollusc shell.

Hypotonic (Gr. *hypo*, under + *tonos*, tension) When the salinity, osmotic pressure, of two solutions is compared, the one with the lower salinity is hypoosmotic to the other and referred to as being hypotonic.

Incomplete gut A digestive system that has only a mouth and no anal opening. Both ingested food and the undigested food must pass through the same opening to the alimentary tract.

Incomplete metamorphosis (Gr. *meta*, between + *morphosis*, to shape or form) Insects that lack a pupal stage in the life cycle. Adults resemble the larval stages in appearance but have wings and a functional reproductive system.

Indirect development When a larval stage hatches from the egg and has little or no resemblance to the adult. Larvae, often free-swimming in aquatic environments, grow and undergo a transformation into the adult form.

Integument (L. *in*, in + *tegere*, to cover) The body surface or covering of an animal.

Intercostal muscle (L. *inter*, between + *costa*, rib) Muscle between the ribs involved, in part, in breathing.

Intermediate host In parasites with complex life cycles involving more than one host, the organisms that contain the larval stages of the parasite.

Intertidal zone Part of the ocean ecosystem covered with water at high tide and exposed at low tide.

Intracellular digestion (L. *intra*, within + *cella*, cell or chamber) The final breakdown of ingested food to components small enough to be absorbed across the cell membrane; occurs inside of the cells in the digestive tract. Digestive enzymes may be secreted for initial breakdown of the meal but phagocytosis ultimately occurs and nutrients are absorbed across the membrane of the phagosome into the cell.

Introvert (L. *intro*, inward or toward the inside + *vertere*, to turn) An eversible proboscis found in a number of the pseudocoelomate phyla. When the pharynx is everted, the hooks on the introvert are used to penetrate and hook onto the substrate or prey.

Isotonic (Gr. *isos*, equal + *tonos*, tension) When the salinity, osmotic pressure, of two solutions is compared and are the same, the solutions are isotonic to each other.

Jacobson's organ (Ludvig Jacobson, 1783-1843) Blind-ended sacs in the roof of the mouth of some reptiles, especially snakes and lizards, used for olfaction, smell. There is no connection to the respiratory system. The tongue is used to transfer sensory chemicals to the surface of Jacobson's organ.

Jaw Made of either cartilage or bone, the jaw is a modified gill arch used for feeding. It may be armed with teeth or hardened plates, and it forms a part of the vertebrate mouth.

Keratinized (Gr. *keras*, horn) Keratin is a protein with high sulfur content from the covalent sulfur bridges that cross-link adjacent proteins. This gives vertebrate hair, nails, feathers, and claws their durability.

Kinetic skull (Gr. *kinetos*, moving) Skull found in squamate reptiles, lizards and snakes; birds; and some fishes. Unlike other skulls, the upper jaw and palate that form the snout or nose can move relative to other parts of the skull.

Kinetoplast (Gr. *kinein*, to move) Specialized mitochondria found at the base of the flagellum of some protozoans.

Kinetosome (Gr. *kinein*, to move + *soma*, body) Structure that forms and anchors the axoneme of a cilium or flagellum in the plasma membrane; also called a basal body.

Labium (L. *labium*, lip) The second maxilla in some uniramians, such as insects, are often fused to form the lower portion of the buccal cavity.

Lacunae (L. *lacunae*, space) Empty spaces or cavities found between other cells that are otherwise solid or continuous.

Lappets Paired flaps of tissue surrounding the sensory rhopalia and arranged around the edge of scyphozoan medusae.

Larval amplification A form of asexual reproduction where, during the life cycle, a single larval organism can produce large numbers of the next developmental stage. A good example is the fluke life cycle. A single sporocyst develops into hundreds of redia.

Lateral line Line along the side of fishes and aquatic amphibians that includes underlying canals and sense organs that detect sound waves and movements in water.

Lemnisci (L. *lemniscus*, ribbon) Pair of epidermal pouches in an acanthocephalan that use hydrostatic pressure to extend the proboscis; pulled back in using muscles.

Leuconoid The most complex of the three different sponge architectures. Choanocytes are found in chambers, and there is no spongocoel. Water enters incurrent canals to the prosopyles and exits the chambers through apopyles, then excurrent canals and the osculum.

Ligament sac Possible remnant of the missing gut in acanthocephalans. The connective tissue of the ligament sac supports the reproductive tract and may divide the body into separate compartments.

Lobopod (L. *lobus*, lobe + Gr, *podus*, foot) Type of pseudopod consisting of both ectoplasm and endoplasm; large, thick, and blunt-ended.

Lophophorates (Gr. *lophos*, crest + *phorein*, to carry) Group of animals that have a lophophore--ring of hollow ciliated tentacles surrounding the mouth.

Lophophore (Gr. *lophos*, crest + *phorein*, to carry) A unique double ring of hollow ciliated tentacles that surround the oral opening in a number of animal phyla. There is some debate as to whether these should be separate phyla, and many consider the lophophore to be a unique character of the phylum Lophophorata.

Lophotrochozoa A new taxon supported by molecular evidence; includes all the animals with a lophophore or a trochophore larval stage. Lophotrochozoans and ecdysozoans, animals with a cuticle, no epidermal ciliature, and who molt, are combined as the protostomes in traditional evolutionary classification of animals.

Lorica (L. *lorica*, corset or girdle) Any external protective casing that surrounds an invertebrate. Loricas are made of either materials secreted by the organism or surrounding materials cemented together to form the protective girdle.

Lunule In sand dollars, notches or holes that pass through the body of the animal. Their function is not fully understood but may be involved in movement of food from the oral to aboral surface, or they may stabilize the animal in strong water currents.

Macrobenthic (Gr. *makros*, large + *benthos*, deep) Larger than 0.5 millimeters and found on or near the bottom of aquatic environments.

Macronucleus (Gr. *makros*, large + *nucleus*, nucleus or nut) One of two types of dimorphic nuclei found in ciliate protozoans. The macronucleus contains multiple copies of the genome (polyploid) and is responsible for general protozoan cell function. The other type of nucleus is the micronucleus.

Madreporite (Ital. *madre*, mother + Gr. *poros*, pore) The sievelike external opening to the water vascular system in echinoderms.

Malpighian tubules (Marcello Malphigi, 1628-1694) Excretory structures found in insects and some spiders. Although similar in appearance and function the two are not homologous. Malpighian tubules are hollow extensions of the gut suspended in the hemocoel.

Mammary gland (L. glandulae mammariae) The milk producing glands that, along with hairs, are characteristic of mammals.

Mandible (L. *mandibula*, jaw) The lower jaw of a vertebrate or the feeding appendages of uniramians and crustaceans.

Mantle (L. *mantellum*, clock or mantle) A thin sheetlike membranous extension of the visceral mass of molluscs and brachiopods that forms two flaps of skin. The mantle secretes the shell on the dorsal side and the space between the two flaps of skin is referred to as the mantle cavity.

Mantle cavity (L. *mantellum*, clock or mantle) In molluscs the mantle secretes a shell that extends body wall, creating a cavity between the mantle and the body of the animal.

Manubrium (L. *manubrium*, handle) Tubular extension of the subumbrellar (oral) surface of a jellyfish medusa; the mouth is located at the tip of the manubrium.

Masseter One of two muscles that close the vertebrate jaw; the other is the temporalis.

Mastax The unique feeding structure of rotifers. It is a modification of the pharynx that includes internal jaws (trophi) used to grind and tear ingested food.

Maxilla (L. dim of *mala*, jaw) Head appendages involved in feeding. They most commonly are involved in manipulating the food so that it can be crushed or chewed by the mandibles.

Maxillae (L. dim of *mala*, jaw) Head appendages involved in feeding. They most commonly are involved in manipulating the food so that it can be crushed or chewed by the mandibles.

Maxilliped (L. dim of *mala*, jaw + *pedis*, foot) Arthropod appendages are referred to as maxillipeds when the first appendages found on the thorax become involved in feeding.

Medusa (Gr. *Medusa*, creature in Greek mythology) The free-swimming, mobile stage of the cnidarian life cycle. This stage, when present, is reproductive and mature gonads form on either male or female medusae. A common example is the jellyfish.

Mehlis' gland Glandular cells surrounding the ootype in trematodes and cestodes. Its role is unclear. It may be involved in formation of the eggshell, lubricating the egg as it moves from to ootype into the uterus, or its secretions may activate the sperm for fertilization.

Meiobenthic (Gr. *meion*, less or smaller + *benthos*, deep) Smaller than 0.5 millimeters and found on or near the bottom of aquatic environments.

Meiofauna (Gr. *meion*, less or smaller + *Fauna*, mythical god of herdsmen) Animals smaller than 0.5 millimeters found on or near the bottom of aquatic environments.

Meiosis (Gr. *meiosis*, reduction or to make small) Cell divisions that result in the formation of gametes--egg or sperm. Meiosis involves two rounds of nuclear divisions; one round involving DNA replication. Each of the four resulting products is haploid and has only one of the paired sets of chromosomes found in a diploid cell.

Mesenchyme (Gr. *meso*, middle + *enchyma*, infusion) The middle layer between the inner and outer cells of an animal with only two cell layers. It forms from ectoderm, consists of a jellylike mesoglea, and may also contain cells.

Mesocoel (Gr. *meso*, middle + *koilos*, hollow) The middle of three coelomic spaces found in the tripartite body plan characteristic of the deuterostome lineage of animals. The other coelomic compartments are the protocoel and metacoel.

Mesoderm (Gr. *meso*, middle + *derma*, skin) The third cell layer that develops in the gastrula between the ectoderm and endoderm in triploblastic animals. Mesoderm develops into muscle, connective tissues, and bones, as well as blood and other components of the vascular system.

Mesoglea (Gr. *meso*, middle + *gloia*, glue) The jellylike layer found between the ectodermal and endodermal cell layers of diploblastic organisms. It acts as a type of cement holding the two layers together but, unlike mesenchyme, has few, if any, cells.

Mesohyl (Gr. *meso*, middle + *hyle*, wood or matter) Gelatinous matrix between the two layers of cells in the body wall of the sponge. Contains spicules and amebocytes but is not homologous with mesoderm.

Mesonephric (Gr. *meso*, middle + *nephros*, kidney) Replaces the pronephros during development and is retained as the functional kidney in adult fish and amphibians. Formed posterior to the pronephros from renal tubules, which ultimately connect with the archinephric duct.

Mesosoma (Gr. *meso*, middle + *soma*, body) In animals where the abdomen (or in chelicerates, the opisthosoma) is divided into two regions, the most anterior is the mesosoma and the posterior the metasoma.

Mesosome (Gr. *meso*, middle + *soma*, body) In the tripartite coelom of deuterostomes, this is the middle of the three coelomic spaces.

Metabolic waste (Gr. *metabole*, change) The biochemical waste products of metabolism. These most commonly include ammonia, carbon dioxide, and water.

Metacercaria (Gr. *meta*, after, among, or between + *kerkos*, tail + L. *aria*, connected with) Encysted stage of the trematode fluke's life cycle. It is not present in all life cycles but when it is, it follows the cercaria and forms a dormant cyst. This cyst, often in fish, is consumed by the definitive host. The adult stage follows the metacercaria.

Metachronal wave During the coordination of the ciliary movement, bands, or groups of cilia, are at different stages of their beating pattern, and this creates a wavelike appearance to the movement of the cilia on the surface of the organism.

Metacoel (Gr. *meta*, after, among, or between + *koilos*, cavity) The last of three coelomic spaces found in the tripartite body plan characteristic of the deuterostome lineage of animals. The other coelomic compartments are the protocoel and mesocoel.

Metamere (Gr. *meta*, after, among, or between + *meros*, part) In segmented animals that have undergone metamerisation, each of the repeated units is a metamere. Each metamere contains identical structures to the metameres adjacent to it.

Metameric (Gr. *meta*, after, among, or between + *meros*, part) A body plan that consists of a series of identical units, metameres, repeated down the longitudinal axis of the animal. Each metamere contains identical structures to the adjacent metameres.

Metamerism (Gr. *meta*, after, among, or between + *meros*, part) The division of the body into a series of identical units, metameres, repeated down the longitudinal axis of the animal. Each metamere contains identical structures to the adjacent metameres.

Metamorphosis (Gr. *meta*, after, among, or between + *morphosis*, to form or shape) Distinct and marked changes between two stages in the life cycle of an organism. Examples are caterpillars and butterflies or tadpoles and frogs.

Metanephric (Gr. *meta*, after, among, or between + *nephros*, kidney) Most developed of the three different kidneys; formed from the posterior part of the nephric ridge and drained by a ureter. Found in reptiles, birds, and mammals.

Metanephridia (Gr. *meta*, after, among, or between + *nephros*, kidney) An excretory-osmoregulatory organ consisting of a ciliated funnel, nephrostome, connected to tubules that lead to the external nephridiopore. The nephrostome collects coelomic fluid, and all of its contents, to produce the urine.

Metasoma (Gr. *meta*, after, among, or between + *soma*, body) In animals where the abdomen (or in chelicerates, the opisthosoma) is divided into two regions, the more posterior is the metasoma and the anterior the mesosoma.

Metasome (Gr. *meta*, after, among, or between + *soma*, body) In the tripartite coelom of deuterostomes, the posterior of the three coelomic spaces.

Metazoan (Gr. *meta*, after, among, or between + *zoon*, animal) True multicellular organisms that exhibit all the characteristics of animals. They have cells, tissues, and organs.

Micronucleus (Gr. *mikros*, small + *nucleus*, nucleus or nut) One of two types of dimorphic nuclei found in ciliate protozoans. The single micronucleus contains only one copy of the genome and is used during the reproductive cell divisions. The other nuclear type is the macronucleus.

Microtrich (Gr. *mikros*, small + *trix*, hair) Small microvillar projections on the surface of the tapeworm tegument. Microtriches increase the surface area for exchange and may interdigitate with the microvilli of the host's intestine. Plural, microtriches.

Microvilli (Gr. *mikros*, small + L. *villus*, tufts of hair) Small, fingerlike projections of the cell surface that increase the surface area available for exchange.

Miracidium (Gr. *meirakidion*, young person) The ciliated stage that hatches from the egg of a trematode fluke and then invades the first intermediate host, usually a snail, where it develops into a sporocyst.

Mitochondria (Gr. *mitos*, thread + *chondros*, granule) Intracellular organelle where oxidative phosphorylation occurs to produce ATP that powers a cell's activities.

Molting (L. *mutare*, to change) The periodic shedding of the outer body covering of an animal. The term is applied to a variety of things including: the loss of the outer exoskeleton in arthropods, the fur of a mammal, feathers of a bird, or the skin of a reptile.

Monociliate (Gr. *monos*, single or alone + L. *cilium*, eyelash or eyelid) Cells that have flagella and, as a consequence, only one 9+2 organelle on each cell. This differs from polyciliate cells that have numerous cilia (9+2 structures) on the surface of each cell.

Monoecious (Gr. *monos*, single or alone + *oikos*, house) Organisms that have, at some time during their life, both male and female reproductive structures. This is the opposite of dioecious.

Monogamous (Gr. *monos*, single or alone + *gyne*, women) Mating system where the male and female form a pair that may last for the duration of a breeding season or for life.

Monophyletic (Gr. *monos*, single or alone) A group of organisms, including the ancestor to that group, that all share a common evolutionary line of descent. (Compare to polyphyletic.)

Multiple fission A form of cell division where a single parent cell divides and produces more than two daughter cells. When only two daughter cells are produced, the process is referred to as binary fission.

Myoepithelium (Gr. *myos*, muscle + *epi*, on or upon + *thele*, nipple or thin skin) Epidermal cells specialized for contraction. Also called epitheliomuscular cells.

Myomere (Gr. *myos*, muscle) Blocks of segmental muscles found in chordates. The term may also be used to refer to similar blocks of muscle in other animals, but is traditionally used for chordates.

Myoneme (Gr. *myos*, muscle + *nema*, thread) Strands of contractile myofibers found in single cells. These allow the cell, or a portion of the cell, to contract in length and change its shape.

Myotome (Gr. *myos*, muscle + *tomein*, to cut) The embryological blocks of mesoderm in vertebrates that form the myomeres. In some cases myomere and myotome are used interchangeably.

Myriapod, Myriapoda (Gr. *myrios*, countless + *podos*, foot) A taxon within the Uniramia that includes animals that have many pairs of legs, but not those with three pairs--the insect

Nacreous layer (Fr. *nacre*, mother-of- pearl) The innermost of the three layers of a mollusc shell. Also referred to as mother-of-pearl, it consists of thin layers or calcium carbonate crystals (aragonite), continually produced by the mantle surface.

Naupliar eye (Gr. *nauplios*, shellfish) The unique eye found in the larval stage of the crustacean life cycle. This single compound eye is located medially, and with the exception of copepods, barnacles, ostracods, and a few others, it disappears and is replaced with paired eyes in adults.

Nauplius larva (Gr. *nauplios*, shellfish) A larval stage in the crustacean life cycle characterized by the single medial compound eye, the naupliar eye.

Nematocyst (Gr. *nema*, thread + *kystis*, sac or bladder) This organelle is part of the cnidocyte unique to the Cnidarians. It is the stinging, or eversible, portion of the cell, and it can drill into, entangle, and/or stick to potential prey.

Nephridia (Gr. *nephridios*, pertaining to the kidney) A general term referring to the metanephridia and protonephridia.

Nephridiopores (Gr. *nephridios*, pertaining to the kidney + *poros*, pore) External openings to nephridia (protonephridia and metanephridia) in invertebrate excretory systems.

Nephrostome (Gr. *nephros*, kidney + *stoma*, mouth) Open-ciliated funnel of the metanephridium; moves coelomic fluid directly into the metanephridium without ultrafiltration.

Neurochord Dorsal nerve cord or concentration of nerves in hemichordates. It is not clear if the neurochord and notochord are homologous or dissimilar structures.

Neutral buoyancy When the overall density of the organism is the same as the surrounding water. Body tissue is denser than water, and air or oils may be used to achieve neutral buoyancy.

Niche The role that a species has in its surrounding environment. The species' way of life and its combination with abiotoc and biotic defines that species and its role.

Nictitating membrane (L. *nictare*, to wink) Thin membranous eyelid between the outer eyelids and the inner surface of the eye. It moves independent of the other eyelids and is found in all but the mammalian vertebrates.

Notochord (Gr. *noton*, back + *chorde*, string) This ancestral characteristic of the Chordata consists of a stiff cartilaginous rod near the dorsal surface of the animal. It is skeletal and helps support to the body.

Nutritive muscular cell The cells that form the gastrodermis lining the inner cavity of cnidarians. They carry out two functions: the first is to absorb and digest food and the second is in movement or changing shape by contraction of the myoneme portion of the cell.

Ocelli (L.) The simple eye of arthropods consisting of a single lens and optical unit. Ocelli are not usually image forming; they detect levels of light instead.

Octaradial symmetry (Gr. *okto*, eight) Symmetry based on eight; common in some types of corals.

Omnivore (L. *omnis*, all + *vorare*, to swallow or devour) An animal that eats both plant and animal foods.

Omnivorous (L. *omnis*, all + *vorare*, to swallow or devour) Eats both plant and animal foods.

Oncomiracidium (Gr. *onkos*, barbed or hooked + *meirakidion*, young person) The ciliated larval stage of a monogenetic fluke that hatches from the egg, swims to its host, and attaches before transforming into a juvenile fluke.

Oncospheres (Gr. *onkos*, barbed or hooked + *sphaira*, globe or sphere) A larval tapeworm, complete with a hooked scolex, contained inside a membranous envelope. When it is released and free-swimming, it is called a coracidium.

Ootype (Gr. *oon*, egg + *typos*, mold) Part of the oviduct in flatworms with openings to the vitelline gland, Mehlis' gland, and seminal receptacle.

Open circulatory system A circulatory system in which the circulating fluid (blood) flows into vessels or tubes not connected to each other by small capillaries but instead enters into the hemocoel before returning to the heart.

Opercular gill (L. *opercula*, cover or lid) The gill found in bony fish. The operculum is the outer covering of the gill. When it is drawn away from the body, it pulls water inside the buccal cavity across the gills. This motion, in combination with the opening and closing of the mouth, aerates the fish gills.

Opisthaptor (Gr. *opisthen*, behind + *haptein*, to fasten) One of two attachment organs in monogenean flukes. Unlike the prohaptor, the opisthaptor is the principal attachment structure and can consist of hooks, claws, and suckers.

Opisthosoma (Gr. *opisthen*, behind + *soma*, body) The second tagma of the cheliciform body, which either lacks appendages or, if they are present, are involved in gas exchange or silk production.

Oral siphon (L. *os*, mouth) Part of the body of an ascidian that surrounds the mouth. Water is pulled into the siphon chamber through the oral siphon by the action of the cilia lining the pharyngial basket.

Oral surface (L. *os*, mouth) In radially symmetric animals there is no anterior, posterior, left, or right sides to the animal. Instead, we refer to the two sides of the animal by the location of the mouth. In this case the side where the mouth is found.

Organs of Tömösvary Sensory organs located at the base of the antenna in myriapods, but not insects. They're assumed to be sensory because of the multiple nerves that converge in the center of the organ. What they detect isn't clear; humidity, pressure, or sounds are possibilities, with the last being the most probable.

Osphradia (Gr. *osphra*, smell) Sensory structure near the gills and in the incurrent flow of water in the mantle chamber of molluscs; assumed to be important in sensing water quality.

Ossicle (L. *os*, bone) Either the small bonelike structures or plates that form the endodermal skeleton of echinoderms, the needlelike deposits of the shell in early molluscs, or the name of small bones found in the vertebrate inner ear.

Ostia (L. *ostium*, door) An opening into a tube, cavity, or internal space. Examples include the openings in the tubular heart of arthropods, the gills of bivalves, the aquiferous system of sponges, and the oviduct of vertebrates.

Ovaries (L. *ovarium*, ovary) Female reproductive organs that produce eggs.

Oviger legs (L. *ovum*, egg + *gerere*, to carry) The first pair of legs in male sea spiders (Pycnogonids) modified in males for holding and taking care of the eggs prior to their hatching. Females use their oviger legs for grooming.

Oviparous (L. *ovum*, egg + *parere*, to bring forth) When the eggs are laid by the female and the embryo develops and hatches outside of the body of the female.

Oviposit (L. *ovum*, egg + *ponere*, to place) To lay or release the eggs.

Ovoviviparous (L. *ovum*, egg + *vivus*, live + *parere*, to bring forth) Animals in which fertilized eggs undergo development inside the female, but don't receive any nutrients from the parent.

Palp (Gr. *palpus*, stroke or caress) Small, sensory appendages, or feelers, associated with the feeding mouthparts of some invertebrates, most notably arthropods.

Pancreas Part of the vertebrate digestive system that secretes digestive enzymes into the small intestine and produces hormones important in regulating blood sugar levels. The pancreas in squids and related cephalopods is a diverticulum of the digestive gland producing digestive enzymes.

Pandora larva Larval stage in asexual cycle in Cycliophora. It is formed from a small bud that settles on the host.

Papilla (L. *papilla*, nipple) Small nipplelike projection; includes the root papilla that forms mammalian hair and epidermal growths on a variety of surfaces.

Parabronchi The small air tubes in a bird's lungs forming the gas exchange surface. Unlike other vertebrates, air flows across the surface in only one direction.

Parapodia (Gr. *para*, beyond or beside + *podos*, foot) Paired lateral, unjointed appendages of polychaete worms. Parapodia have a variety of shapes, and their appearance is related to their different roles in locomotion and respiration in different polychaete worms.

Parapodium (Gr. *para*, beyond or beside + *podos*, foot) One of the paired lateral, unjointed appendages of polychaete worms. Parapodia have a variety of shapes, and their appearance is related to their different roles in locomotion and respiration.

Parasite (Gr. *para*, beyond or beside + *sitos*, bread, food) Where one animal survives and lives at the expense of another, its host.

Parazoa (Gr. *para*, beyond or beside + *zoon*, animal) Animals organized at the cellular grade. They lack tissues; evidence of this is the absence of a basement membrane or tight cell junctions between the cells. Usually includes Porifera and Placozoa.

Parenchyma (Gr. *para*, beyond or beside + *enchein*, to pour in) Mesenchyme cells that form a spongy mass surrounding the internal organs and other acoelomate invertebrates.

Parthenogenesis, Parthenogenetically (Gr. *parthenos*, virgin + *genesis*, birth or origin) A form of asexual reproduction where viable offspring develop from unfertilized eggs that, depending on the organism, may be either haploid or diploid.

Pectine (L. *pectine*, comb) Modified appendages of the opisthosoma of a scorpion. Although their function is uncertain, they are thought to be mechanoreceptive sensing vibrations in the substrate and chemoreceptive.

Pectoral girdle (L. *pectus*, breast) Bones in vertebrates that connect the appendages on the left and right side of the anterior appendicular skeleton to each other. The pectoral girdles are also attached to the axial skeleton in amphibians, reptiles, birds, and mammals.

Pedalia The flat, enlarged base of the tentacles in some cnidarians, usually the cubozoans.

Pedicellaria (L. *pediculus*, little foot + *aria*, like or connected to) Defensive pincerlike structure found on the surface of echinoderms. They are capable of responding to external stimuli independently from the main nervous system of the animal. Some complex pedicellariae are poisonous.

Pedicle (L. *pedicullus*, little foot) Footlike, stalklike, or stemlike structure.

Pedipalp (L. *pedis*, foot + *palpus*, stroke or caress) The second pair of appendages on the body of a cheliciforme arthropod.

Pedomorphosis (Gr. *pais*, child + *morphe*, to form or shape) When sexually mature adults have characteristics that would normally be associated with the larval stage. It happens because reproductive structures form in the larval before the morphological change to the adult occurs.

Pelagic (Gr. *pelagos*, sea) One of the zones in the marine environments where animals live. Pelagic animals are only found swimming or floating between the surface and bottom of the ocean and not at the shoreline.

Pellicle (L. *pellicula*, skin) The network of semirigid cell membrane thickenings found on the surface of some protozoans. These are used to anchor either the locomotory flagella or cilia into the surrounding plasma membrane.

Pelvic girdle (L. *pelvis*, a basin) Bones in vertebrates that connect the appendages on the left and right side of the posterior appendicular skeleton to each other. The pelvic girdles are also attached to the axial skeleton in amphibians, reptiles, birds, and mammals.

Pentaradiate (Gr. *pente*, five) Radial symmetry based on five. This type of symmetry is unique to the Echinodermata.

Pericardial cavity (Gr. *perikardion*, around the heart) The part of the coelom that forms a sac, or space, that surrounds the heart. In invertebrates with a hemocoel, the pericardial cavity is all that remains of the true coleom.

Periostracum (Gr. *peri*, around, + *ostrakon*, shell) The outermost of the three layers of a mollusc shell. This horny layer consists of chonchin, a protein that protects the underlying layers from damage.

Perisarc (Gr. *peri*, around + *sarx*, flesh) The transparent, chitinous, outer covering of the stalk and branches in a hydroid colony.

Peristomium (Gr. *peri*, around + *stoma*, mouth) The part of the body behind the prostomium in annelids that contains the oral opening. Like the prostomium it is not a true segment and has no setal hairs, which are characteristic of the phylum. (Some textbooks erroneously refer to this as the first true segment.)

Peritrophic membrane (Gr. *peri*, around + *trophe*, nutrient, nourishment) Porous, chitin/protein membrane surrounding food in the midgut of some insects. Presumed roles are to protect the delicate midgut from abrasion by ingested food, a protective barrier against ingested pathogens, and the creation of specialized spaces for different stages of digestion.

Perivisceral cavity (Gr. *peri*, around + L. *viscera*, internal organs) The body cavity that surrounds the main organ systems in an animal.

Perivisceral coelom (Gr. *peri*, around + L. *viscera*, internal organs + Gr. *koilos*, hollow) The coelomic space surrounding the main organ systems in an animal.

Pesticide resistance When the target insects are able to survive a concentration of pesticide that previously was effective at controlling or killing them.

Phagocytosis (Gr. *phagein*, to eat + *kytos*, cavity or cell) Cellular ingestion; engulfing food particles or foreign bodies by surrounding them with cytoplasm and forming a membrane-bound vesicle or vacuole containing the particle.

Pharyngeal basket (L. Gr. *pharynx*, throat) In ascidian adults the pharynx is enlarged to form the pharyngeal basket. Pharyngeal openings in the wall of the basket allow water to pass through with food being trapped on the inner surface before being passed into the gut.

Pharyngeal gill slit (L. Gr. *pharynx*, throat) Lateral opening in the wall of the pharynx that allows water to enter into the mouth and exit through the pharynx. This is one of the ancestral characteristics of the phylum Chordata and are also found in Hemichordata and Enteropneusta. It allowed water to be removed from ingested food before it was passed back into the digestive system.

Pharynx (L. Gr. *pharynx*, throat) The region of the digestive tract between the mouth and esophagus. In most animals it is muscular and forces food into the digestive tract that lies behind it. In vertebrates it is part of both the digestive and respiratory tracts.

Phenetic taxonomy (Gr. *phaneros*, visible or evident) A method of classifying organisms based on numeric methods that evaluate organisms by a variety of measurements to create taxa without any consideration of phylogeny.

Photosynthetic (Gr. *phos*, light + *synthesis*, incorporation, or to put together) Biochemical process that captures light energy to produce simple sugars and oxygen; most often carried out by chloroplasts.

Phyllopodous (Gr. *phyllon*, leaf + *podos*, foot) In primitive crustacea the inner and outer surfaces of the leg are often enlarged to form flaps, or leaflike structures, that the crustacean uses for locomotion, feeding, and respiration.

Phylogeny (Gr. *phyle*, tribe + *genesis*, descent or origin) The evolutionary history of either an organism or a taxon.

Pilidium (Gr. *pilos,* felt cap) Ciliated larval stage of nemertean worms. It has a mouth but no anus and looks like a winter hat with earflaps.

Pinacocyte (Gr. *pinax*, plank or tablet + *kytos*, cavity or cell) Flattened cells that cover the surface of a sponge. Collectively they form the pinacoderm.

Pinacoderm (Gr. *pinax*, plank or tablet + *derma*, skin) Outer layer of pinacocyte cells in a sponge. It is not a tissue, only a layer of cells.

Pineal organ (L. *corpus pineale*) Endocrine gland in vertebrates that produces melatonin, which may be involved in the biological rhythms.

Pinnule (Gr. *pinna*, feather) In crinoid echinoderms the surface area of the arms is increased by smaller side branches, or extensions, referred to as pinnules.

Pinocytosis (Gr. *pinein*, to drink + *kytos*, cavity or cell) Uptake of fluid, rather than particulate material, by endocytosis; surrounds the fluid with cytoplasm to form a vesicle or vacuole.

Placenta (L. *placenta*, a flat cake) In placental mammals, an exchange surface between the unborn fetus and the mother's uterine wall. The fetal and maternal blood supplies remain separate but nutrients, metabolic wastes, and gases are exchanged across the surface.

Placoid (L. *plax*, plate + *eidos*, form) Scales characteristic of the cartilaginous fish. Formed from the dermis, the scale is anchored in that layer by a basal plate composed of dentin and from that a spine, made of toothlike enamel, points backward.

Planktonic (Gr. *planktos*, wandering) Small organisms that float and drift in water.

Planula (L. *planus*, something flat) A planula is the solid, free-swimming larval stage of cnidarians consisting of two cell layers-- an outer ciliated ectoderm and an inner endoderm.

Planuloid (L. *planus*, something flat) Something that resembles a planula, the solid, free-swimming larval stage of cnidarians consisting of two cell layers-- an outer ciliated ectoderm and an inner endoderm.

Plasmalemma Although occasionally used to refer to any cell membrane, a plasmalemma is usually associated with plants and some amebas.

Platyzoa (Gr. *platys*, flat + *zoon*, animals) The infrakingdom of animals that have three tissue layers but lack a coelom. Molecular evidence supports the taxon that includes Platyhelminthes, Gastrotricha, Gnathostomulida, Rotifera, and the recently discovered Cycliophora.

Pneumatized bone (Gr. *pneumo*, wind or air) Bone that contain numerous spaces and sinuses. Found in birds, these bones are light and strong due to internal struts and cross braces.

Polyp (Gr. *polypous*, many footed) The sessile, asexual stage in the cnidarian life cycle. In some species they are independent organisms; in others, they form colonies where some polyps are involved in food gathering (gastrozooids) and other polyps (gonozooids) produce the reproductive stage.

Polyphyletic (Gr. *polys*, many + *phyle*, tribe) A taxon that includes animals from two or more distinct evolutionary lineages and may not include the ancestor of either. (Compare to monophyletic.)

Polypide (Gr. *polypous*, many footed) The soft part of the ectoproct (bryozoan) that includes the visceral mass and the lophophore and can be retracted into the cystid.

Polyploid (Gr. *polypous*, many footed) Cells that have more than the normal haploid chromosomal number.

Porocyte (Gr. *poros*, pore + *kytos*, cavity of cell) A cell in an asconoid sponge with a hole through its center so that water can enter the sponge.

Precaudal (L. *prae*, before + *cauda*, tail) The part of the body in front of the tail.

Preen Used to describe how birds groom and clean their feathers.

Primitive characters The traits, or characteristics, that an animal has that are ancestral to the taxon.

Prismatic The middle of three layers in the typical molluscan shell.

Proboscis (L. *proboscis*, snout or trunk) Tubular feeding structure that extends from an animal. In many invertebrates, it is eversible and extends from the body to permit feeding.

Procercoid (L. Gr. *pro*, before + Gr. *kerkos*, tail) Stage in the aquatic life cycle of tapeworms that develops from the ciliated free-swimming coracidium when it invades a copepod intermediate host. The procercoid stage develops into the pleurocercoid once it is in a fish.

Procuticle (L. Gr. *pro*, before + *cuticula*, skin) Inner layer, under the epicuticle, of the arthropod cuticle. Composed of chitin and protein, it may be differentiated into two layers in terrestrial arthropods and three in the crustacea.

Proglottids (L. Gr. *pro*, before + *glotta*, tongue) Serially repeating segmentlike structures found in tapeworms that contain the reproductive organs. Immature proglottids are continually added at the scolex, and the most mature proglottids, containing thousands of eggs, are found the farthest from the scolex.

Prohaptor (L. Gr. *pro*, before + *haptein*, to fasten) The specialized paired attachment suckers on either side of the mouth in monogenean flukes. It consists of adhesive structures that help attach the animal to the host. The opisthaptor, not the prohaptor, is the principal attachment structure.

Prokaryote (L. Gr. *pro*, before + *karyon*, nut or nucleus) Cells or organisms with cells, that don't have a membrane-bound nucleus.

Pronephric (L. Gr. *pro*, before + *nephros*, kidney) The first, or ancestral, kidney that appears in the anterior part of the coelomic cavity and is connected to the archinephric duct. In amniotes and bony fish it appears only in the early stages of the embryo before it disappears.

Prosoma (L. Gr. *pro*, before + *soma*, body) The first tagma of a cheliciform consisting of the first six segments of the body. Appendages on the prosoma are involved in feeding and locomotion. The prosoma has no sensory appendages such as antennae.

Prosome (Gr. *protos*, first + *soma*, body) Anterior most of the three body cavities, or regions, in deuterostome animals.

Prostomium (L. Gr. *pro*, before + *stoma*, mouth) The most anterior part of an annelid found just in front of the mouth it. Like the peristomium, it is not a true segment.

Protocerebrum (Gr. *protos*, first + L. *cerebrum*, brain) The first part of the arthropod brain that includes the optic lobes and the nerves from the compound eyes.

Protocoel (Gr. *protos*, first + *koilos*, hollow) The first of three coelomic spaces found in the tripartite body plan characteristic of the deuterostome lineage of animals. The other coelomic compartments are the mesocoel and metacoel.

Protonephridia (Gr. *protos*, first + *nephros*, kidney) Osmoregulatory-excretory structure found in some invertebrates. Also called a flame-cell, this tubule is closed at its distal end. The beating of internal cilia pull water across the cell membrane and then propels it down the tubule.

Protonymphon (Gr. *protos*, first + *nymphe*, bride, chrysalis) The unique larval stage of sea spiders, pycnogonids. The protonymphon has only three pairs of appendages.

Protopodite (Gr. *protos*, first + *podos*, foot) The basal part of the crustacean biramous appendage that attaches it to the body. Consists of two parts: coxa and basis.

Protostome (Gr. *protos*, first + *stoma*, mouth) Phyla that share common characteristics of the blastopore forming the mouth, spiral determinate cellular cleavage in the embryo, and the formation of the body cavity by schizocoely.

Prototroch (Gr. *protos*, first + *trochos*, wheel) Ciliated band around the middle of the trochophore larva; just anterior to the mouth.

Protozoans (Gr. *protos*, first + *zoon*, animal) Single-celled organisms, in the kingdom Protista, that exhibit the animal-like characteristic of having to feed to obtain nutrients. They are considered heterotrophic.

Pseudocoelom (Gr. *pseudes*, false + *koiloma*, cavity) A body cavity that is not completely lined by mesoderm. The mesoderm is associated with only the ectodermal surface but not the endoderm. There is some debate on how the psedocoelom forms. It is by definition a remnant of the blastocoel but often body cavities are defined as pseudocoeloms without properly making this developmental link.

Pseudocoelomate (Gr. *pseudes*, false + *koiloma*, cavity) Animals that have a body cavity that is not completely lined by mesoderm. In the past these organisms were referred to as the phylum Aschelminthes but this is no longer considered acceptable because there is no apparent common ancestor to the group.

Pseudopod (Gr. *pseudes*, false + *podos*, foot) A cytoplasmic extension that extends from the surface of either a protozoan or any amoeboid cell. These structures are temporary and are used for locomotion and feeding.

Pulmonary system (L. *pulmo*, lung) That part of the circulatory system the transports blood from the heart to the lung, or other respiratory surface, and back to the heart. It's counterpart is the systemic system.

Pygidium (Gr. *pyge*, rump or tail + *idion*, ending) The most posterior part of a segmented animal that houses the anal opening. The pygidium is not a true segment.

Quadraradial symmetry (L. *quattuor*, four) Body symmetry based on four identical parts arranged around a central axis.

Radial cleavage (L. *radius*, ray or spoke) During development as the cells of the zygote divide, the products of the cell division remain stacked directly on top of each other.

Radial symmetry, Radially symmetric (L. *radius*, ray or spoke) When an organism's body parts are arranged around the oral-aboral axis so that any plane passing through this axis results in two identical halves.

Radula (L. *radula*, scrapper) This unique feeding structure is an ancestral characteristic of all animals in the phylum Mollusca. It looks like a tongue covered with teeth and works like a file to rasp food off the substrate.

Ram ventillation Movement of water across the surface of the gills by swimming and holding the mouth open.

Redia, Rediae (Francesco Redi, 1626-1698) A stage in the life cycle of trematode flukes. The redia in the intermediate host develop from germ cells in the sporocyst. Redia also contain numerous germ cells that will develop in cercaria, an example of larval amplification.

Renette cell Unique cells found in nematodes believed to be involved in osmoregulation and elimination of metabolic wastes.

Repugnatorial glands (L. *repugnare*, to resist) Defensive glands that secrete offensive secretions. They're found in millipedes and some spiders.

Respiratory tree (L. *respiratio*, breathing) The unique respiratory structure found in sea cucumbers (Echinodermata). Water is pumped in and out of these structures found inside the body cavity.

Reticulopodia (L. *reticulum*, net + Gr. *podos*, foot) Type of pseudopod consisting of thin cytoplasmic strands that form a mesh or net.

Retrograde (L. *retro*, backward + *gradi*, step) Moving toward the back.

Rhabdite (Gr. *rhabdos*, rod) Found in the epithelium of free-living flatworms, and when released dissolve to produce mucus, which may also help in defense because it contains toxic chemicals.

Rhabdoids (Gr. *rhabdos*, rod) Rodlike structures produced by the epithelium of free-living turbellarian flatworms.

Rhopalia (Gr. *rhopalon*, club) These sensory structures are found around the bell margin of the jellyfish medusa. They always contain a statocyst for balance and sometimes an ocelli for light detection.

Rhynchocoel (Gr. *rhynchos*, snout + *koilos*, hollow) The unique body cavity of the Nemertean worms, which contains the retractable proboscis. There is some debate on whether this is a true coelom, making it difficult to state whether the Nemertea are eucolomates.

Saprophytic (Gr. *sapros*, rotten) Feeding on dead or decaying organic materials.

Scalid Rings of spines at the base of the head directed toward the posterior end of the body of Kinorhynca, Priapulida, and Loricifera.

Schistosomiasis (Gr. *schizein*, to cleave or split + *soma*, body) A disease caused by the trematode parasite schistostoma. Different species are responsible for diseases ranging from common swimmers itch to schistosomiasis, which effects over 200 million people.

Schizocoel (Gr. *schizein*, to cleave or split + *koilos*, hollow) A true body cavity that forms by schizocoelus splitting of the mesoderm.

Schizocoelus (Gr. *schizein*, to cleave or split + *koilos*, hollow) One of two ways that a body cavity forms within the mesoderm. (The other is enterocoelic pouching.) The block of mesoderm splits apart to create the coelomic space, a characteristic of the protostomes.

Schizogony (Gr. *schizein*, to cleave or split + *gone*, birth) A form of asexual reproduction found in some protozoans. An already multinucleated cell undergoes cell division that results in each daughter cell containing only one of the many nuclei present in the parent cell. This is also referred to as multiple fission.

Sclerites (Gr. *skleros*, hard) Hardened plates; usually of arthropod cuticle.

Sclerocytes (Gr. *skleros*, hard + *kytos*, cavity or cell) Cells that produce spicules in sponges.

Sclerotized (Gr. *skleros*, hard) Chemical covalent cross-linking of separate protein chains using phenolic compounds. Sclerotized protein is stable and is not easily broken down or digested.

Scolex (Gr. *skolex*, worm or grub) This unique attachment organ of the tapeworms is the most anterior part of the animal and is used to attach to the host. It consists of adhesive suckers and, in some species, hooks. Proglottids develop from behind the scolex.

Scyphistoma (Gr. *skyphos*, cup + *stoma*, mouth) The jellyfish, class Scyphozoa, get their name from this unique stage in the life cycle. It is a small polyp existing for only for a short time before developing into a strobila, which then produces the medusa.

Scyphomedusa (Gr. *skyphos*, cup + *Medusa*, creature in Greek mythology) The medusa of scyphozoan Cnidaria.

Scyphostome (Gr. *skyphos*, cup + *stoma*, mouth) The polyp stage in the life cycle of scyphozoan Cnidaria; formed from the planula after it settles and undergoes transverse fission to form the strobila and the small medusoid ephyra.

Sebaceous gland (L. *sebum*, tallow or L. *glandula sebacea*) Mammalian epidermal gland cell that produces the oily sebum used to maintain hairs.

Secondary palate (L. *secundus*, second + L. *palatum*) Plate of bones that separates the nasal and oral cavity in mammals. Also found in some reptiles.

Segmentation (L. *segmentum*) The division of the body into a series of identical segments, metameres, that are repeated down the longitudinal axis of the animal. Each metamere contains identical structures to the metameres adjacent.

Semicircular canal (L. *half* + *circulus*, circle) Part of the inner ear consisting of three semicircular ducts that are arranged at right angles to each other. They are sensitive to position and angular acceleration of the head in the three different axes of space.

Seminal receptacle (L. *semen*, seed) Part of the female reproductive system where sperm received from the male during mating is stored and later used to fertilize the eggs.

Seminal vesicle (L. *semen*, seed + *vesica*, bladder) In invertebrates, part of the male reproductive tract where sperm is stored prior to mating; the accessory gland in the male reproductive system of amniote vertebrates that produces the fluid part of the ejaculate.

Septum (L. *septum*, wall) A sheet of tissue that separates two compartments or cavities. Plural, septa.

Serial homology, Serially homologous (Gr. *homologos*, agreeing or corresponding) Metamerization results in a linear series of segments that share a common embryonic origin. Ancestrally, all metameres were identical. Modification of different parts of the body for different functions means that this similarity has been lost.

Setae (L. *seta*, bristle) Bristles, or hairlike structures, that extend from the body. Usually made of chitin, they are common in annelids and arthropods.

Sexually dimorphic (Gr. *dis*, two + *morphe*, shape) When the sexes in a species are physically different in appearance.

Simple eye Also referred to as an eye spot or ocellus; consists of photoreceptive cells positioned underneath a lens that can either change its shape or position to focus light. May be image-forming on some (nautilus and spiders) or only detect light intensities.

Siphonoglyph (Gr. *siphon*, tube or pipe + *glyphein*, to engrave or carve) Single or paired ciliated grooves in the oral opening of anthozoans that, along with other features, create the biradial symmetry of the group. The cilia propel water into the gastrovascular cavity.

Siphuncle (Gr. *siphon*, tube or pipe) A cord of tissue that passes between the different chambers of a Nautilid shell. The siphuncle regulates the amount of air or fluid in the chambers to assure neutral buoyancy for the animal as it swims.

Sister group Pair of taxa that are the closest phylogenetic relatives of each other.

Skeletal girdles In vertebrates, bones that connect the appendages on the left and right side of the appendicular skeleton to each other. The skeletal girdles are also attached to the axial skeleton in amphibians, reptiles, birds, and mammals.

Solenia The channels running through the mass of mesoglea, coenenchyme, that connect the polyps of some corals.

Spermatheca (Gr. *sperma*, seed or sperm + *theca*, case or cover) Part of the insect female reproductive system where sperm received from the male during mating is stored and later used to fertilize the eggs.

Spermatophore (Gr. *sperma*, seed or sprem + *pherein*, to carry) Sperm package passed to the female or inserted into the female reproductive tract by the male.

Spherical symmetry (Gr. *sphaira*, ball or sphere) Symmetry based on a sphere.

Spicules (L. *spiculum*, small spike) Any needlelike structure. This term is most often thought of in conjunction with sponges and refers to the needlelike structures produced by sponge cells that form the supporting skeleton. In nematodes spicules are used by the male during copulation. Needlelike deposits of the shell in Mollusca are also referred to as spicules.

Spinneret The modified appendages on the posterior end of the opisthosoma of a spider that produce silk.

Spiracle (L. *spiraculum*, air hole) Openings to the tracheal system of uniramians and some spiders; also a small or vestigial gill opening in some cartilaginous fishes.

Spiral cleavage (L. *spiralis*, coil or spiral) Pattern of cell division in the developing embryo where the products of the cell divisions shift by rotating either clockwise or counterclockwise so that the resulting daughter cells lie in the furrow of the underlying pair of cell. The opposite of radial cleavage.

Spiralia (L. *spiralis*, coil or spiral) Taxon of animals with spiral embryonic cleavage of the cells during early development.

Spiral valve (L. *spiralis*, coil or spiral) In sharks, a modification of the digestive tract to slow the movement of food and to increase the surface area. In an amphibian, it helps to keep systemic and pulmonary blood separate in the single ventricle of the heart.

Spongin (Gr. *spongos*, sponge) Collagenlike material that forms the skeletal system of some sponges.

Spongocoel (Gr. *spongos*, sponge + *koilos*, hollow) The internal cavity of asconoid and syconoid sponges that opens to the outside through the osculum. There is no spongocoel in a leuconoid sponge.

Sporocyst (Gr. *sporos*, seed + *kystis*, bladder or pouch) A stage in the life cycle of trematode flukes. The sporocyst develops from the mericidium found in the intermediate host. Each sporocyst contains the germ cells that will develop into numerous rediae, an example of larval amplification.

Sporogony (Gr. *sporos*, seed + *gonos*, birth) A form of asexual reproduction where the fusion product of the male and female gamete (the zygote) undergoes multiple cell divisions that produce sporozites. Found in animals in the protozoan phylum Apicomplexa.

Squalene Oil produced by sharks. It assists with maintaining neutral buoyancy.

Stapes Auditory bone in the middle ear of mammals.

Statoblast (Gr. *statos*, standing + *blastos*, bud) Asexual bud of freshwater bryozoans (ectoprocts) capable of surviving unfavorable conditions.

Statocyst (Gr. *statos*, standing + *kystis*, bladder or pouch) A balance organ that senses gravity. It consists of at least one solid statolith surrounded by sensory cilia. As the position of the organism changes, the statolith rolls stimulating different cilia.

Sternal keel (Gr. *sternon*, breast or chest) Modified sternum in birds; the attachment of the flight muscles.

Sternite (Gr. *sternon*, breast or chest) Hardened plate of cuticle on the ventral surface of an arthropod segment.

Sternum (Gr. *sternon*, breast or chest) Breast bone or bone forming the middle wall of the chest in most terrestrial vertebrates; cuticular plate on the ventral surface of an arthropod segment.

Stolon (L. *stolo*, shoot or branch) In some colonial cnidarians, ascidians, pterobranchs, and lophophorans, a cord of tissue connects different zooids of the colony to each other. Buds on the stolon give rise to new zooids, which retain their connection to the rest of the colony through the stolon.

Strobila (Gr. *strobilos*, anything twisted or a cone) Stage in the life cycle of scyphozoan cnidarian formed from transverse fission of the scyphostome and producing the small medusoid ephyra.

Strobilization (Gr. *strobilos*, anything twisted or a cone) The process that converts the scyphistoma into a strobila during the life cycle of scyphozoans, jellyfish. Transverse divisions of the strobila produce the small, disc-shaped ephyra that develop into the adult jellyfish.

Style sac Houses the rotating crystalline style in the mollusc stomach. Cilia lining the style sac rotate the style that releases digestive enzymes into the stomach as it winds the mucus string of food into the stomach.

Subesophageal (L. *sub*, under or below + Gr. *oissein*, to carry + *phagein*, to eat) Below the esophagus.

Suctorial feeders (L. *sugere*, to suck) In this type of feeding strategy food is sucked into the digestive tract by the muscular action of the pharynx.

Suspension feeding Feeding by removing particulate food suspended in the surrounding water.

Swim bladder Found in bony fish, this gas-filled chamber is used to maintain neutral buoyancy. Oxygen in the blood is added or removed as required. The swim bladder in some fishes opens into the digestive system allowing these fish to swallow air instead.

Syconoid (Gr. *sykon*, fig) Of the different sponge architectures, this is intermediate in its complexity. The spongocoel is no longer lined with choanocytes, now located in radial canals that open to the spongocoel through apopyles. Water enters the radial canals through prosopyles and exits through a single osculum.

Symbiotic (Gr. *symbioun*, to live together) Two or more organisms dependent on each other for survival; each organism benefits in some way from the association.

Synapomorphy (Gr. *syn*, with or together + *apo*, from + *morphe*, form) A new and unique character that a group of organisms all share and which defines the lineage or clade.

Synapsid (Gr. *syn*, with or together + *apsis*, arch) Vertebrate skull of the mammals and the extinct mammal-like reptiles. There is only one temporal opening behind the orbits in the skull.

Syncytium, Syncytial (Gr. *syn*, with or together + *kytos*, cavity or cell) Protoplasm that contains numerous nuclei not separated from each other by plasma membrane. This creates a multinucleate cellular appearance for a tissue that appears to lack cell boundaries.

Syrinx (Gr. *syrinx*, pipe) Sound box of a bird; located at the lower end of the trachea.

Systemic system That part of the circulatory system that transports blood from the heart out to the body and back to the heart. Its counterpart is the pulmonary system.

Tagmata (Gr. *tagama*, order, row, or corps) The distinct body regions resulting when different segments of a metameric animal become involved in specific functions. These segments are modified to carry out that function and their appearance changes. Singular, tagma

Tagmatization (Gr. *tagama*, order, row, or corps) The distinct body regions resulting when different segments of a metameric animal become involved in specific functions. These segments are modified to carry out that function and their appearance changes.

Tagmosis (Gr. *tagama*, order, row, or corps) The distinct body regions resulting when different segments of a metameric animal become involved in specific functions. These segments are modified to carry out that function and their appearance changes.

Taxa (Gr. *taxis*, arrangement) Any taxanomic group of related organisms. Singular, taxon.

Tegmen (L. *tegere*, to cover) The outer epithelium covering a crinoid.

Tegmentum (L. *tegumentum*, from L. *tegere*, to cover) Middle layer of the chiton shell composed of organic material, calcium salts, and colored pigments.

Tegument (L. *tegumentum*, from L. *tegere*, to cover) The outer covering of parasitic flatworms, including flukes and tapeworms, consisting of a syncytial outer layer of cytoplasm connected to cell bodies embedded deep in the underlying protective mesenchyme.

Telotroch (Gr. *telos*, end + *trochos*, wheel) Ciliated tuft surrounding the anus of a trochophore larva.

Telson (Gr. *telson*, extremity) In a number of arthropods the most posterior part of the body, which may appear as a tagma, is not a true segment. This part of the body, formed from the embryonic pygidium, is referred to as a telson.

Tergite (L. *terga*, back) Hardened plate of cuticle on the dorsal surface of an arthropod segment.

Test (L. *testum*, crucible or shell) An outer, nonliving shell or case that surrounds an organism.

Testes (L. *testis*, witness) Male reproductive organs that produce sperm.

Tetrapod (L. *trados*, four + *podos*, foot) Vertebrates that have four limbs, or appendages. The forward pair is attached to the pectoral girdle and the posterior pair to the pelvic girdles. Tetrapods include amphibians, reptiles, birds, and mammals.

Thorax (Gr. *thorax*, chest) In organisms that have undergone tagmosis the trunk often develops into two tagma, one involved in locomotion and one that is not locomotory. The locomotory tagma is referred to as a thorax.

Tissue grade Animals that have tissues but no organ systems. Cnidaria and Ctenophora are examples.

Tornaria larva (L. *Tornare*, turn) Larval stage of enteropneusts (hemichordates). It revolves as it swims and bears a resemblance to the larval stages of some echinoderms.

Torsion An unusual twisting of the gastropod body that has left members of the class with an asymmetric body plan and a U-shaped alimentary tract, with the normally posterior anus and mantle cavity now located anteriorly over the head. Some gastropods have modified this further and detorted changing the position of the anus.

Totipotent (L. *totus*, whole, or all + *potentia*, ability or potency) When a differentiated cell can change into any one of a variety of different specialized cells found in an organism. These changes will result in the cell carrying out new functions.

Trachea (Gr. *tracheia*, rough) Part of the respiratory system in insects, vertebrates, and some spiders. In the invertebrates it consists of tubules carrying air directly from outside to the tissues. In vertebrates tracheae are tubes moving air to the lungs.

Tracheal system The respiratory system found in insects and some spiders where air passes directly to the tissues through a series of tubules. Because oxygen transport is direct, respiratory pigments are often missing from the blood of animals that use a tracheal system.

Transverse fission Asexual reproduction by fission at a right angle to the main axis of the body. The strobila in scyphozoan cnidarians uses transverse fission to produce the small medusoid ephyra.

Trichocysts (Gr. *trix*, hair + *kystis*, pouch or bladder) Organelle fired by some ciliates in defense; also used to anchor the ciliate in place.

Tripartite body plan (L. *tres* or *tria*, three) A characteristic of the deuterostome line of animals the coelom forms with three separate compartments referred to as the protocoel, mesocoel, and metacoel.

Tripartite coelom (L. *tres* or *tria*, three + Gr. *koiloma*, cavity) Characteristic of the deuterostome line of animals. The coelom forms with three separate compartments referred to as the protocoel, mesocoel, and metacoel.

Triploblastic (L. *tres* or *tria*, three + *blastos*, germ or bud) Organisms formed from the three cell layers: endoderm, ectoderm, and mesoderm.

Triradiate (L. *tres* or *tria*, three) Radial symmetry based on three.

Tritocerebrum (Gr. *tritos*, three + L. *cerebrum*, brain) The third part of the arthropod brain; in Crustacea, is innervated by the second pair of antennae. It also connects the brain to the subesophageal ganglion.

Trochophore (Gr. *trochos*, wheel + *pherein*, to bear) This free-swimming ciliated larval stage is found in a number of animal phyla including the Mollusca and Annelida. The larvae has a unique circle of preoral cilia around the middle of the body. Trochophores are often considered an ancestral characteristic of protostomes.

Trophi (Gr. *trophe*, nourishment, nutrition) The paired chitinous jaws found inside the mastax of a rotifer. The trophi are adapted to the feeding strategy of the rotifer and can grind, cut, or be used to capture prey.

Trophosome (Gr. *trophe*, nourishment, nutrition + *soma*, body) Remnant of the pogonophoran gut that contains chemosynthetic symbiotic bacteria.

Trunk In animals with only more that one body regions, or tagmata, the trunk can be either the part of the body behind the head, the torso which doesn't include the limbs; or the combined thoracic and abdominal tagma.

Tube foot Hollow and fluid-filled tube that are part of the water vascular system in Echinoderms. Muscles associated with the tube foot allow it to be hydraulically controlled and function in locomotion, attachment, food gathering, and gas exchange.

Tunic (L. *tunica*, tunic or integument) The outer covering of the tunicates secreted by the underlying body wall.

Tunicin Protein that forms the tunic of tunicates, an early group of chordates.

Tuns The dormant and resistant stage of tardigrads.

Tympanum Auditory membrane of the ear; thin cuticle that detects sounds in insects.

Typhlosole (Gr. *typhlos*, blind + *solen*, chamber or pipe) This invagination of the gut wall in earthworms increases the surface area of the gut available for digestion.

Undulipodia Generic term for cilia and flagella.

Uniramous (L. *unus*, one + *ramus*, branch) A type of appendage in arthropods. Uniramous appendages are unbranched and each element of the leg is arranged in a linear sequence.

Urea (Gr. *ouron*, urine) Form of nitrogenous waste with intermediate toxicity between ammonia and uric acid.

Ureters (Gr. *oureter*, ureter) Tubes that carry urine produced in the kidney to the bladder in mammals or to the cloaca in birds and reptiles.

Uric acid (Gr. *ouron*, urine) Nitrogenous waste of birds, reptiles, insects, and some spiders. Uric acid is the least toxic of the three primary nitrogenous wastes, which also include ammonia and urea.

Urns Multicellular structures in the sipunculid body cavity presumed to trap and sequester wastes and debris from dead cells.

Urogenital (Gr. *ouron*, urine + L. *gignere*, to beget) Combination of the excretory and reproductive systems.

Uropygial gland (Gr. *oura*, tail + *pyge*, rump) Oil gland used by birds to maintain their feather when they groom.

Uterine bell Part of the female reproductive system of an acanthocephalan. The uterine bell sorts developing embryos inside the pseudocoel and allows only those that have developed to the right stage to pass into the uterus and out the gonopore.

Vas deferens (L. *vasa*, vessel) Sperm duct that carries sperm away from the testes.

Vector (L. *vehere*, to carry) The animal (often an insect) that carries a disease from an infected individual to one that is not. For example, the mosquito is the vector of malaria in humans.

Velarium (L. *velarium*, awning) Extension of the sides of the subumbrellar surface of cubozoan medusa.

Veliger (L. *velum*, veil or covering) The ciliated larval stage after the trochophore in the life cycle of many molluscs. Its structures include the initial shell, mantle, visceral mass, head, and foot.

Velum (L. *velum*, veil or covering) A thin flap of tissue found around the inner surface of the bell of a hydrozoans medusa.

Ventral (L. *venter*, belly) In animals with bilateral symmetry, this is the lower surface of the animal if it crawls across the substrate and the front of the animal if it walks upright.

Ventricle (L. *ventriculus*) Chamber of the heart that pumps the blood out and away from the heart.

Vermiform (L. *vermis*, worm) Animals with a wormlike appearance with a cylindrical body, considerably longer than it is wide.

Vibrissae (L. *vibrissae*, nostril hair) The whiskers or stiff, tactile, sensory hairs surrounding the nose and mouth of many mammals.

Visceral mass (L. *viscera*, internal organs) One of three parts of a mollusc: head, foot, and visceral mass. It consists of the fleshy part of the mollusc and contains the main organ systems of the mollusc.

Vitelline glands (L. *vitellus*, yolk) Gland in many invertebrates that provision the egg with yolk, which supplies the developing embryo with nutrients.

Viviparous (L. *vivus*, living, + *parere*, to give birth) Animals in which fertilized eggs undergo development inside the female and receive additional nutrients from the parent.

Water expulsion vesicle Special organelles found in protozoan and some parazoans that is involved in osmoregulation. The organelle collects water from the cytoplasm and then releases it from the cell. They are also called contractile vacuoles.

Water vascular system A characteristic of echinoderms. It is a modification of the coelom, and this closed, water-filled system forms canals and branches throughout the body. In one part, the tube feet, it acts as hydrostatic skeleton permitting locomotion.

Wheel organ Complex ciliary bands located in front of the oral opening of cephalochordates that propel food into buccal cavity.

Yolk Nutrient in an egg; rich in protein and fats and used as a food reserve by the developing embryo.

Yolk gland Gland in many invertebrates that provision the egg with yolk, which supplies the developing embryo with nutrients.

Zoacia The nonliving outer exoskeleton of an ectoproct (bryozoan) colony. Singular, zoacium.

Zoaria Colonies of lophophorate ectoprocts (bryozoans). Singular, zoarium

Zoecium (Gr. *zoon*, animal + *oikos*, house) The nonliving outer exoskeleton of an ectoproct (bryozoan) colony. Plural, zoacia

Zonite The name for the thirteen segments of a kinorhynch. The segments are triangular in cross section. Rings of spines at the edges of the zonite help in movement.

Zooid (Gr. *zoon* animal) In colonial animals, each member of the colony. Often there is a division of labor between zooids, the most common being between feeding gastrozooids and reproduction gonozooids.